Iceland Within the Northern Atlantic 1

SCIENCES

Geoscience, Field Director – Yves Lagabrielle

Lithosphere-Asthenosphere Interactions, Subject Head – René Maury

Iceland Within the Northern Atlantic 1

Geodynamics and Tectonics

Coordinated by

Brigitte Van Vliet-Lanoë

iSTE

WILEY

First published 2021 in Great Britain and the United States by ISTE Ltd and John Wiley & Sons, Inc.

ISTE Ltd
27-37 St George's Road
London SW19 4EU
UK

www.iste.co.uk

John Wiley & Sons, Inc.
111 River Street
Hoboken, NJ 07030
USA

www.wiley.com

Library of Congress Control Number: 2021932010

British Library Cataloguing-in-Publication Data
A CIP record for this book is available from the British Library
ISBN 978-1-78945-014-9

ERC code:
PE10 Earth System Science
PE10_5 Geology, tectonics, volcanology
PE10_13 Physical geography
PE10_18 Cryosphere, dynamics of snow and ice cover, sea ice, permafrosts and ice sheets

MIX
Paper from responsible sources
FSC® C013604

Contents

List of Abbreviations ix

Preface xv
Brigitte VAN VLIET-LANOË and Françoise BERGERAT

Introduction xix
Brigitte VAN VLIET-LANOË and René MAURY

Chapter 1. Iceland, in the Lineage of Two Oceans 1
Brigitte VAN VLIET-LANOË and Françoise BERGERAT

1.1. Geographic and geodynamic context. 5
1.2. Components of the North Atlantic domain 7
1.2.1. The Mid-Atlantic Ridge 7
1.2.2. The North Atlantic Igneous Province 7
1.2.3. The Icelandic hot spot 8
1.2.4. The Greenland–Iceland–Faroe Ridge 11
1.3. Geodynamic characteristics of Iceland. 11
1.3.1. Seismicity 12
1.3.2. Icelandic volcanism. 13
1.3.3. Eustatism and the Icelandic glaciers 14
1.4. References 18

Chapter 2. Iceland, an Emerging Ocean Rift 21
Françoise BERGERAT

2.1. Mid-Atlantic Ridge and Icelandic hot spot interactions 21
2.2. Present-day deformations in Iceland 23
2.2.1. Seismicity 24
2.2.2. Motions at plate boundaries 39
2.3. Iceland's main structural features 40
2.3.1. The paleo-rifts and the active rift 41
2.3.2. The transform zones 58
2.4. Geothermal energy and hydrothermalism 86
2.4.1. Geothermal systems 86
2.4.2. Geysers and hydrothermalism 94
2.5. References 97

Chapter 3. Iceland, A legacy of North Atlantic History 115
Laurent GEOFFROY

3.1. Bathymetry of the Northeast Atlantic domain and geoid anomalies . . 115
3.2. The North Atlantic and the continental breakup of Laurussia 117
3.2.1. Passive margins and large igneous provinces 118
3.2.2. The early beginnings of the opening of the North Atlantic Ocean 125
3.2.3.Thulean magmatism in the Paleocene and the continental breakup of the Northeast Atlantic 133
3.2.4. Chronology and kinematics of the opening of the Northeast Atlantic 142
3.2.5. The Northeast Atlantic region: mantle plume or not? 151
3.3. The origin of Iceland 159
3.3.1. The anomalous crust of the GIFR ridge and the deep structure of Iceland 159
3.3.2. Icelandic SDRs 162
3.3.3. Interpretations of GIFR and Iceland 165
3.4. References 170

Conclusion 179
Françoise BERGERAT and Laurent GEOFFROY

References 185

List of Authors 199

Index 201

Summary of Volume 2 207

List of Abbreviations

σ_1, σ_2 and σ_3 Maximum, intermediate and minimum principal stresses of the stress tensor

σ_{Hmax} Maximum horizontal stress

A

AMO Atlantic multidecadal oscillation

B

BTVP British Tertiary Volcanic Province

C

CGFZ Charlie–Gibbs Fracture Zone

CGPS Communicative Global Positioning System

D

DL Dalvik Line

DMM Depleted MORB mantle

DO Dansgaard–Oeschger event

Iceland Within the Northern Atlantic 1,
coordinated by Brigitte VAN VLIET-LANOË.

DSOW	Denmark Strait overflow water
DTM	Digital terrain model

E

E-MORB	Enriched mid-ocean ridge basalts
EM	Enriched mantle
EUR	Europe
EVZ	East Volcanic Zone

F

FLF	Flat-lying flows

G

GEBCO	General Bathymetric Chart of the Oceans
GIA	Glacio-isostatic adjustment
GIFR	Greenland–Iceland–Faroe Ridge
GIR	Greenland–Iceland Ridge
GL	Grimsey Line
GPS	Global Positioning System

H

HFF	Húsavík-Flatey Fault
HIMU	High Mu mantle (Mu = U/Pb)

I

ICPMS	Inductively coupled plasma mass spectrometry
IFR	Iceland–Faroe Ridge
IGS	International GPS Service
IMO	Icelandic Meteorological Office (Veðurstofa Íslands)
InSAR	Interferometric Synthetic Aperture Radar
IRD	Ice-rafted detritus
ISOW	Iceland–Scotland Overflow Water
ÍSNET	GPS Network surveys of the National Land Survey of Iceland (Landmælingar Íslands)

J

JMFZ	Jan Mayen Fracture Zone

K

KR	Kolbeinsey Ridge

L

LBA	Labrador–Baffin axis
LGM	Last Glacial Maximum (extension)
LIP	Large igneous provinces

M

M or M_W	Moment magnitude
MAR	Mid-Atlantic Ridge

M_b	Body-wave magnitude
M_L	Local magnitude
M_S	Surface-wave magnitude

N

N-MORB	Normal mid-ocean ridge basalts (depleted)
NADW	North Atlantic Deep Water
NAIP	North Atlantic Igneous Province
NAM	North America
NEIC	National Earthquake Information Center (United States)
NGRIP	North Greenland Ice Core Project
NVZ	North Volcanic Zone

O

OIB	Ocean island basalts
OSC	Overlapping spreading center

R

RP	Reykjanes Peninsula
RR	Reykjanes Ridge

S

SDRs	Seaward-dipping reflectors
SIL	South Iceland Lowland network
SISZ	South Iceland Seismic Zone

T

TFZ	Tjörnes Fracture Zone

U

USGS	United States Geological Survey

W

WVZ	West Volcanic Zone

Preface

Brigitte VAN VLIET-LANOË and Françoise BERGERAT

This collective work is the logical conclusion of more than 30 years of French research in Iceland, with the support of various programs and institutions. It has also benefitted from the contribution of a CNRS Thematic School on Iceland, which was held in Brest in 2010 and which was strongly impacted by the eruption of Eyjafjallajökull. This book is the fruit of the work of a group of complementary researchers who are very fond of Iceland. Our thoughts turn to Jacques Angelier who left this basaltic ship a little too early. There are multiple authors to each chapter – with a principal author for each one – in order to provide a multidisciplinary approach to the discussed scientific problems and take into account all our publications up to the most recent ones (2019–2020).

French research in Iceland began in the mid-1980s, initiated by Françoise Bergerat (Sorbonne Université, formerly Université Pierre et Marie Curie, in Paris) in search of an "emerging oceanic ridge", in collaboration with Jacques Angelier†, then Catherine Homberg. Very quickly, this collaboration was extended to Icelandic colleagues, Águst Guðmundsson (London), Kristjan Sæmundsson, Ragnar Stefánsson and Sigurdur Rögnvaldsson †. The first work focused on the analysis of brittle deformations and then turned to sismotectonics.

Iceland Within the Northern Atlantic 1,
coordinated by Brigitte VAN VLIET-LANOË. © ISTE Ltd 2021.

This work was then supplemented, from the 2000s, by the geodetic campaigns of the team from the Université de Savoie in Chambéry led by Thierry Villemin in collaboration with Halldór Geirsson and his group. At the beginning of the 1990s, Laurent Geoffroy began (in Paris) work on the Thule basaltic provinces (Scotland, Ireland, Faroe Islands), continued from the 2000s (at the Université du Maine, in Le Mans) on the other side of the Atlantic, in Greenland. The analysis of the morphology of Iceland began in the mid-1990s at the Université de Rennes-I, with Olivier Dauteuil and Brigitte Van Vliet-Lanoë, and then extended to the neighboring ocean in relation to volcanism and the evolution of the North Atlantic. At the same time, the Neogene and Quaternary climatic history of the island, recorded by stratigraphy, was consolidated with dating carried out by Hervé Guillou and his colleagues and by geochemistry carried out at the Université de Bretagne Occidentale, in Brest, in close collaboration with Águst Guðmundsson (Hafnafjördur), Kristjan Sæmundsson and Helgi Björnsson's team. The last stage of this work is currently being developed in the Géosciences Océan laboratory in Brest, with Laurent Geoffroy and René Maury. It concerns the evolution of the North Atlantic based on Icelandic and Greenlandic data.

The material and logistical support of the Icelandic authorities proved to be very constructive both for field work and for data acquisition and sharing: IMO (Veðurstofa Íslands/Icelandic Meteorological Office); ISOR (Íslenskar orku-rannsóknir/Icelandic energy research), formerly Orkustofnun (National Energy Authority); Landsvirkjun (National Power Company) and Vatnajökull National Park. This research would not have been as fruitful without the physical and intellectual help of all our students, at Master's level and/or with their thesis works: Olivier Bourgeois, Magalie Bellou, Jean-Christophe Embry, Loïc Fourel, Sebastian Garcia, Guillaume Gosselin, Solène Guégan, Romain Plateaux, Lionel Sonnette, Anne Sophie Van Cauwenberge, Ségolène Verrier and Audrey Wayolle.

Finally, this work was made possible because of the assistance of the French Embassy in Iceland and funding from the European Commission, 4th and 5th PCRD (PRENLAB-1 and -2, PREPARED and SMSITES programs); the Paul Émile Victor Institute (IPEV), formerly the Institut français pour la recherche et la technologie polaires (IFRTP) (Arctic Program 316); the Icelandic Ministry of Education; and the French Ministry of Foreign Affairs (Franco-Icelandic scientific and cultural collaboration program).

We also thank Bernadette Coleno, Marion Jaud, Laurent Gernigon and Alexandre Lethiers for their contributions to the figures in this volume.

February 2021

Introduction

Brigitte VAN VLIET-LANOË and René MAURY

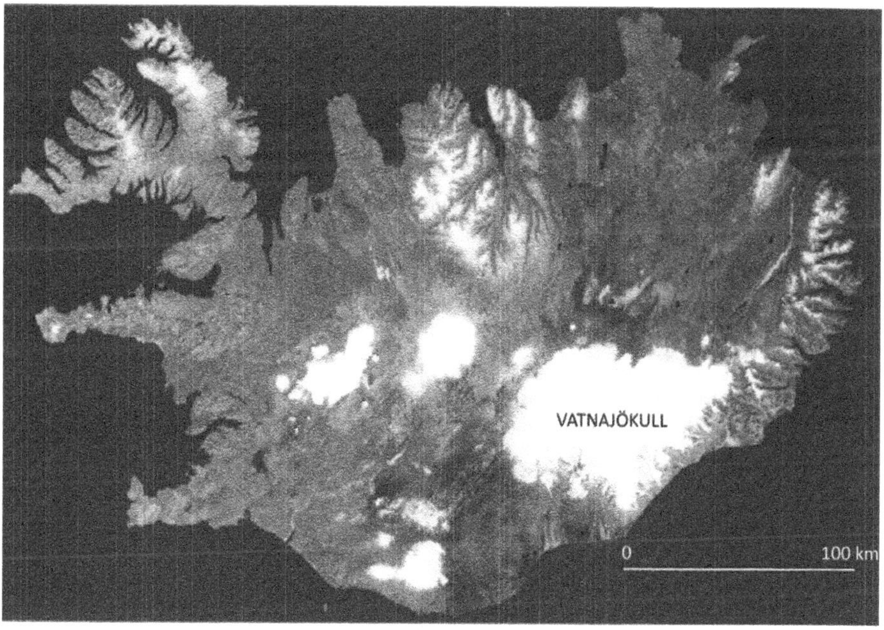

Figure I.1. *Iceland from Space (document Geographical Institute of Iceland/Landmælingar Ísland [LMIs])*

For color versions of the figures in this Introduction see, www.iste.co.uk/vanvliet/iceland1.zip.

Iceland Within the Northern Atlantic 1,
coordinated by Brigitte VAN VLIET-LANOË.

Iceland (Figure I.1), a young and isolated island in the middle of the Atlantic Ocean, has only very recently been discovered in terms of the scale of human history. Irish monks (the papars) passed from island to island in their curraghs (Figure I.2) via the Shetland Islands and the Faroe Islands to evangelize the legendary Hyperborea. These journeys took place as early as the 6th century, a period with cold volcanic winters.

The papars discovered a world of fire and ice, the gates of hell. They settled in round peat-covered huts and dug shovel caves in the consolidated sandy interglacial formations in the south of the island.

Figure I.2. *(A) Icelandic stamp illustrating the discovery of Iceland by Irish monks (papars). (B) The glacial lake Jökulsarlón dominated by the volcano Oræfajökull (Brigitte Van Vliet-Lanoë©)*

Two hundred years later, the Vikings, warriors but also more than anything farmers in search of cultivable land (Figure I.3), settled in the south and west of the island from 860 AD on wooded land made fertile by thick layers of volcanic loess. This is the *landmana* of the Icelandic sagas. They installed their parliament, the Alþing, around 900 AD, in a remarkable site (Figure I.4), which became a high place of plate tectonics, the Þingvellir graben, the boundary between the European and American plates.

These fertile lands were surmounted for at least 400,000 years by a fire monster, the Hekla volcano (Figure I.5). Its Plinian eruption in 1104 AD (H1, volcanic explosivity index of 5) destroyed many Viking settlements in the Rangavellir, not only by falling pumice and gas but also by the associated glacial megafloods, the *jökulhlaups*, submerging the Þjorsárdalur with a wave of muddy water more than 25 m high. At that time, the Hekla must have been more ice-covered than it is today.

Figure I.3. *Traditional sheepfolds in southern Iceland with the volcano Hekla in the background (Rangavellir) (Brigitte Van Vliet-Lanoë©)*

Figure I.4. *(A) The Þingvellir graben, seen from an airplane (Þingvellir National Park Web site). The Alþing site is located in front of the white buildings (chapel). (B) Reconstruction of the Alþing in the Middle Ages by W.G. Collingwood (1897)*

Figure I.5. *(A) The Hekla volcano (Brigitte Van Vliet-Lanoë©) and (B) its cartographic representation on the map Islandia of Ortelius (1585)*

Despite the island's long isolation from continental Europe, there is a lot of information about its history. Indeed, Icelanders have jealously preserved their language and ancient books, including the famous sagas, and have often proved to be great writers and avid readers, even on isolated farms.

In addition to their literary and historical interest, the sagas represent a source of exceptional paleo-environmental information on a period whose climatic evolution was very complex: the Medieval Optimum and the climatic degradation that followed. The University of Iceland was founded in 1911 and, due to its special nature, Iceland is the country with the highest proportion of geologists and especially volcanologists among its population.

Figure I.6. *Typical landscape of ancient basalts on the eastern coast of Iceland (Skridalur) (Brigitte Van Vliet-Lanoë©)*

Iceland is a land of fire and ice, still sparsely populated (about 350,000 inhabitants in 2020), prized by tourists for its "unspoiled", photogenic character and its many natural wonders, although Viking colonization quickly made the forest disappear. But recent tourist development has also caused an invasion of 4×4 vehicles, brand new hotels and vacation huts, raising the standard of living of the population, but gradually destroying a natural heritage – including the geological heritage – surprisingly well preserved until the early 21st century. Industrial development (geothermal, hydroelectricity and electrometallurgy) kept the population in the peripheral sectors of the island and above all modified the landscape of the coastal zones. Whatever one does or looks at in Iceland is *de facto* connected to the geological history of the island (Figure I.6).

Despite its remoteness, Iceland is a land that directly influences Western Europe through its position in the north-central Atlantic, as a beacon of the Gulf Stream and thermohaline circulation, or through its meteorological depression. But it is also a land consisting mainly of layered basaltic piles, still active from a tectonic and volcanic point of view. We were reminded of it by the last eruption of the Eyjafjallajökull (March–October 2010) with its plume of ash that invaded Europe and disrupted intercontinental commercial flights (Figures I.7 and I.8).

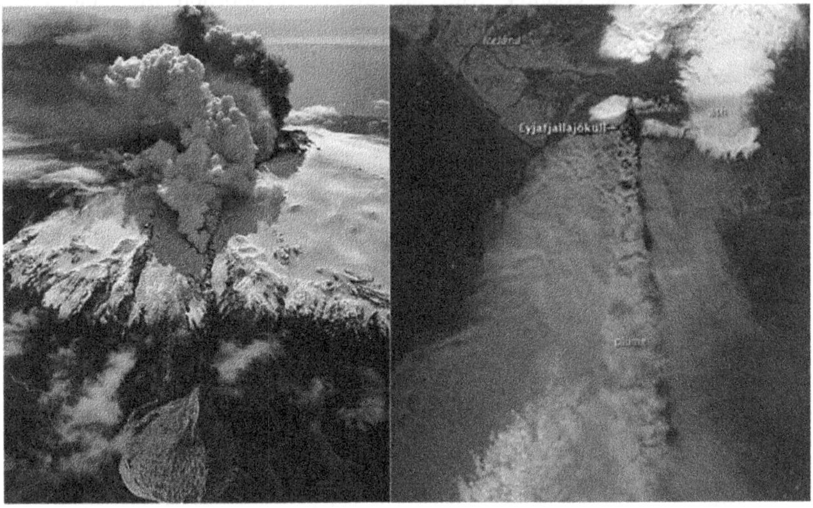

Figure I.7. *The eruption of the Eyjafjallajökull in 2010: its jökulhlaup (Reykjavik Helicopter©) and plume (Earth Observatory, NASA)*

Figure I.8. *Sheep disturbed by the ash from the eruption (Flickr©)*

The glaciers are located on volcanic edifices, considered to be at least Quaternary. The largest ice cap, the Vatnajökull, rests on some of the most active volcanoes of the island, located above the summit of a deep magma plume.

Bárðarbunga (Figure I.9) is one of the volcanoes found above the Icelandic hotspot and is located on the western margin of the present Vatnajökull ice cap.

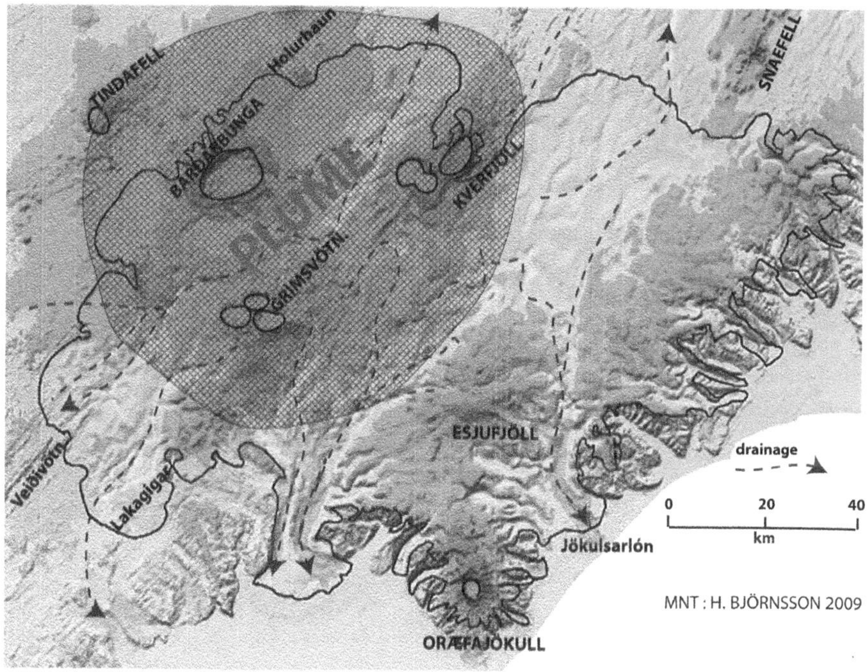

Figure I.9. *Digital terrain model of Vatnajökull (black line: current cap boundary) completed with the flood drainage positioning and the potential extension (to a depth of 200 km) of the Icelandic mantle plume (in gray) (source: H. Björnsson, 2009)*

The most recent eruption of this volcano (August 2014–February 2015; Figure I.10) was linked to the draining of a magma chamber located 12 km below the caldera, following the climate driven melting of the cap (about 1 m/year).

To the northwest, the most impressive lava flow since the 18th century, the Holhurhaun flow, occurred along a fracture line, in association with swarms of earthquakes that stretched to the Askja volcano in the north. The previous eruption, that of Veiðivötn, had flown toward the south in 1747, awakening the Torfa volcano at the same time.

The Bárðarbunga is also a source for *jökulhlaups* or megafloods, resulting from the melting of the glaciers by the heat of the lava emitted and which mostly flow toward the north.

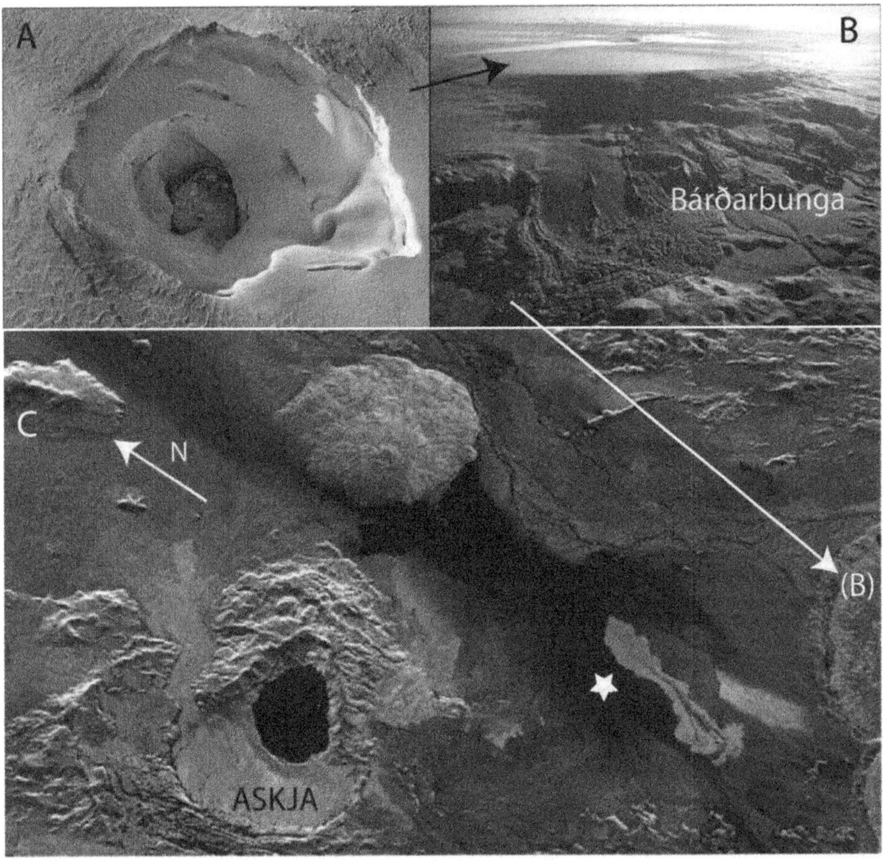

Figure I.10. *(A and B) Views of the Bárðarbunga caldera obliterated by ice during the 2014 eruption with a melting cauldron to the west (A and black arrow) (photo: mbl.is/RAX). (C) Satellite image of the emersion of the Holhurhaun flow (star) on August 13, 2014 at the foot of the Bárðarbunga (B) (photo: TerraSAR-X)*

Another major volcanic structure is located in the center of the ice cap, directly above the top of the mantle plume: it is the triple caldera of Grímsvötn (Figures I.9 and I.11), which emitted the vast majority of basaltic tephra that hide the glaciers and reach the lands surrounding the North Atlantic.

The most famous is the Saksunarvatn tephra splayed around 10,200 years cal BP. This volcano is never at rest; its current eruptive frequency is about 10 years and it also remained continuously active during the Ice Age, but with a lower frequency.

It is mainly responsible for the formation of subglacial lakes and is at the origin of most of the *jökulhlaups* that gully the emissaries of the Vatnajökull cap (Figures I.12 and I.13). At present, these floods mainly destroy road infrastructures such as the Main Highway (N1).

Figure I.11. *Rim of the northern caldera of Grìmsvötn (Ragnar Sigurdsson©)*

In northern Iceland, volcanic activity is also significant, in association with the northern rift. Many geothermal fields are exploited there, such as the Krafla field northeast of Lake Myvatn (Figures I.14 and I.15).

This volcanic activity also occurs at sea, both in the north in the Kolbeinsey Ridge and its intermittent island (white point in Figure I.16(A)) and in the southwest along the Reykjanes Ridge, or in the Vestmann Islands, a southern extension of the East Volcanic Zone.

Figure I.12. *The Grìmsvötn volcano. (A) Initiation of the northward flow associated with a collapse of the ice mass (sun to the west), which led to the great jökulhlaup of November 1996 (Oddur Sigurðsson©). (B) Grimsvötn crater at the end of the 2011 eruption (Dima Moiseenko©). (C) Interstratified and deformed basaltic tephras in the terminal glacier tongue of Brúarjökull (LMIs)*

Figure I.13. *(A) The jökulhlaups of the Skafta River from the Grimsvötn in 1996 (M.T. Gudmundsson©) and (B) Main Highway (N1) in 2011 (Veðurstofa Íslands©), frequently repaired since 1970, with (C) the jökulhlaup memorial of November 1996: two enormous pieces of the metal deck of the old bridge, twisted like common wires (Françoise Bergerat©)*

Figure I.14. *(A) Fissural eruption of Krafla in 1980, along fractures arranged en échelon. (B and C) The geothermal power plant (C) narrowly escaped destruction by lava flows (flow with white arrow) (B-C: Brigitte Van Vliet-Lanoë©)*

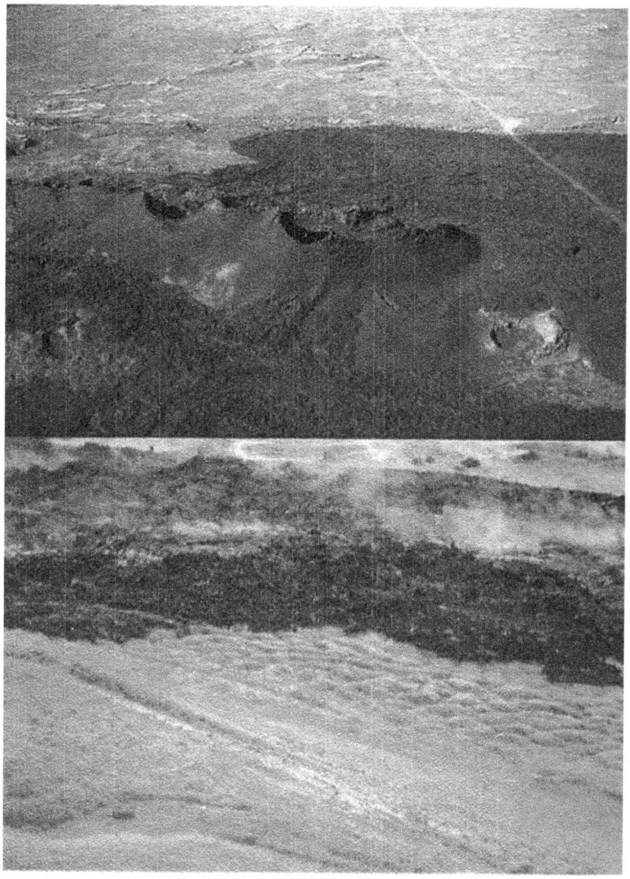

Figure I.15. *Eruption of Krafla in 1980: hornitos on fractures and lava flows in 1997 (Brigitte Van Vliet-Lanoë©)*

The latter were the locus of a first submarine eruption in 1963 (building of the Surtsey volcano), then of a fissural eruption (followed by a strombolian phase) partially destroying the town of Vestmanayer on the main island of Heimaey in 1973.

In relation to volcanic activity and especially to tectonic activity, seismicity is permanent in Iceland and major earthquakes have regularly occurred, particularly in the northern peninsulas (Húsavík region) and in the whole south of the island. When crossing the lava fields between Hveragerði and the Hekla, many remarkably preserved traces of major historical earthquakes (M > 6) can be observed (Figure I.17).

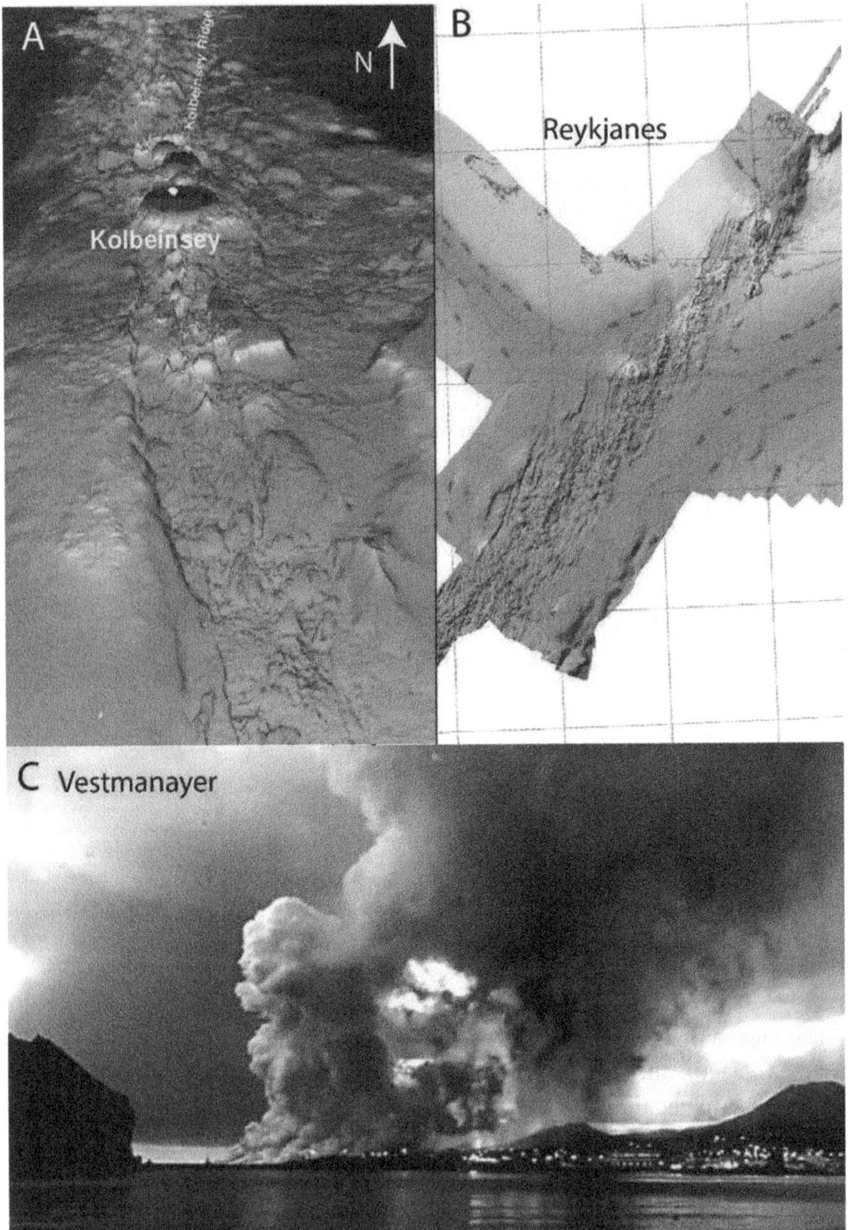

Figure I.16. *(A) The submarine ridges of Kolbeinsey and (B) of Reykjanes (multibeam echosounder images, HAFRO.is). (C) The eruption of Heimaey, building the "Mountain of Fire" (Eldfell) in 1973*

Figure I.17. *Trace of the Réttarnes seismic fault (1294 or 1732) in the Rangavellir: South Iceland Seismic Zone (Françoise Bergerat©)*

If in the north of the island the current earthquakes occur mainly offshore, the Húsavík and Kopasker agglomerations are however far from being sheltered from a significant seismic event, and in the south, several major earthquakes have occurred very recently (M_w 6–7; June 2000, May 2008).

While Icelandic houses are relatively insensitive to earthquakes (Figure I.18), the same cannot be said for road infrastructure or greenhouses. The temporary rise or fall of water tables or lakes is frequent, reactivating or deactivating geysers and causing fluid escapes. This is particularly the case in the Hveragerði or Geysir region: Strokkur is currently the most active and Great Geysir is currently intermittent (Figure I.19).

The cold pole of Iceland is represented by its glaciers, currently relatively little extended but which covered practically all the island at the time of the last glaciation, inhibiting the activity of a great majority of the volcanoes. Most of the time, they settled at the top of the volcanic edifices constituting the high points of the island such as Vatnajökull (2,009 m at Bárðarbunga) or Hofsjökull (1,765 m at Habunga; Figure I.20).

Figure I.18. *Destruction caused during the earthquakes of June 2000 in Bitra: (A and B) farm buildings, (C) main highway (N1) (A-B-C Françoise Bergerat©), and at the end of May 2008 in Hveragerði: (D) dislocated pipes and damaged greenhouses (Brigitte Van Vliet-Lanoë©)*

Figure I.19. *Successive phases of an explosion of the Strokkur geyser, Geysir geothermal field (Brigitte Van Vliet-Lanoë©)*

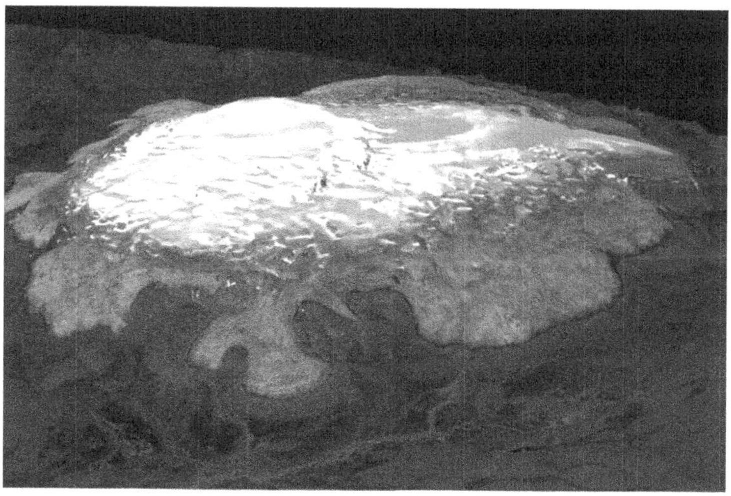

Figure I.20. *The Hofsjökull. Document made from radar images (CNES©). The caldera is located at the top left of the picture*

These glaciers have profoundly carved the island since the Neogene, with deep glacial valleys, ice-smoothed or striated rocks, countless drumlins and large areas of abandoned glacial sediments on the central plateau, especially around the Kerlingarfjöll (Figure I.21). Some volcanoes have typically subglacial morphologies, such as tabular volcanoes or tuyas, the best known of which is Herðubreið (Figure I.22). Others form alignments of ridges, the *tindar*, which formed at the margin of the melting caps (Figure I.23).

The waters from these glaciers have also shaped canyons with huge waterfalls, on powerful, gray and loaded water rivers, the *jökullsá* (Figures I.23 and I.24). These waters are currently collected for an important hydroelectric production with mainly industrial purposes (aluminum and rare metals extracted from imported ores). This resource accounts for 72% of Iceland's electricity production.

Various cap outlets are currently being developed and managed, with water stored in very large dams, generally superimposed on the same course and designed to resist *jökulhlaups* of interglacial rank. Global warming in recent decades and potentially induced volcanism are likely to call this policy into question.

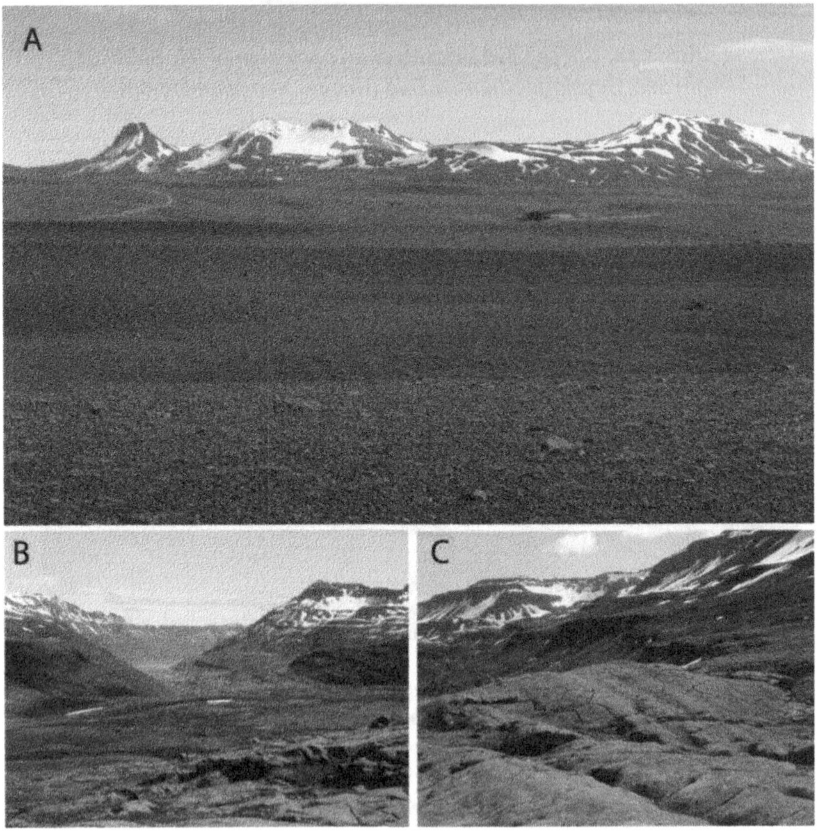

Figure I.21. *(A) The Kerlingarfjöll surrounded by its glacial desert. (B) Perched upper cirque and (C) ice-smoothed rocks of the eastern fjords (Mjóifjörður, south of Seiðifjörður) (Brigitte Van Vliet-Lanoë©)*

Figure I.22. *A subglacial tabular volcano: the Herðubreið, north of Vatnajökull, North Volcanic Zone (Brigitte Van Vliet-Lanoë©)*

Figure I.23. *Jökulsá á Kreppa north of Vatnajökull with hyaloclastite or tindar ridges (Brigitte Van Vliet-Lanoë©)*

Figure I.24. *A key Icelandic resource: water. (A) Bruarjökull outlet (the glacier is at the bottom of the photograph) (LMIs©). (B) One of the Dettifoss waterfalls (Jökulsá á Fjöllum). (C) The Haslsón dam on the Jökulsá á Brú. (D) The Fannahlið aluminum plant (Hvalfjörður) (photos B, C and D: Brigitte Van Vliet-Lanoë©)*

On land, Iceland's only important and renewable resources are its water and, as a result, its hydroelectricity, as well as its many geothermal sites related to the presence of the hot spot.

In this two-volume book, we will present the geological and glacial history of this island, its current tectonic and volcanic activity and the impact of its formation on the climatic evolution of the last few millions of years. Volume 1 replaces Iceland within the geological framework of the North Atlantic, and describes its tectonic and geodynamic evolution. Volume 2 (Van Vliet-Lanoë 2021) is dedicated to the study of the interactions between Icelandic volcanism and external geodynamics, i.e. with glaciations and the climatic evolution of the Atlantic zone during the Neogene and the Quaternary.

1

Iceland, in the Lineage of Two Oceans

Brigitte VAN VLIET-LANOË and Françoise BERGERAT with the collaboration of René MAURY, Hervé GUILLOU and Laurent GEOFFROY

There is no active region on Earth comparable to Iceland and its ocean environment (Figure 1.1). Situated in the Northeast Atlantic, Iceland is indeed located within a particularly complex geodynamic evolutionary domain, illustrating the problems inherent to the break-up of continents in the plate tectonic model, especially the recycling of ancient structures such as the suture of the Iapetus Ocean (section 3.2.2.3) and its paleoslab active during the Silurian. The latter has been mapped along the eastern coast of Greenland and is associated with calc-alkaline palaeo-volcanism (Andresen *et al.* 2007; Rhenström 2010).

The existence of Iceland probably dates back to the Oligocene. Its position east of Greenland and its insularization make it a key witness of the great changes controlling the evolution of the oceanic circulation. It has thus controlled the evolution of the climate since the Neogene, through the North Atlantic Current. It has also been, since 9 My, a key recorder of the onset of glaciation in the northern hemisphere, and also of glacier–volcanism interactions. Its location in the middle of the Greenland–Faroe Islands Ridge (GFIR) controls the descent of cold salty waters from the Arctic and North Atlantic into the thermohaline circulation (section 3.4 of Volume 2).

For color versions of the figures in this chapter see www.iste.co.uk/vanvliet/iceland1.zip.

Iceland Within the Northern Atlantic 1,

Figure 1.1 summarizes the broad physiographic features and toponymy of the different structural domains (ridges and basins) of the North Atlantic and the Labrador-Baffin axis (LBA) on both sides of Greenland.

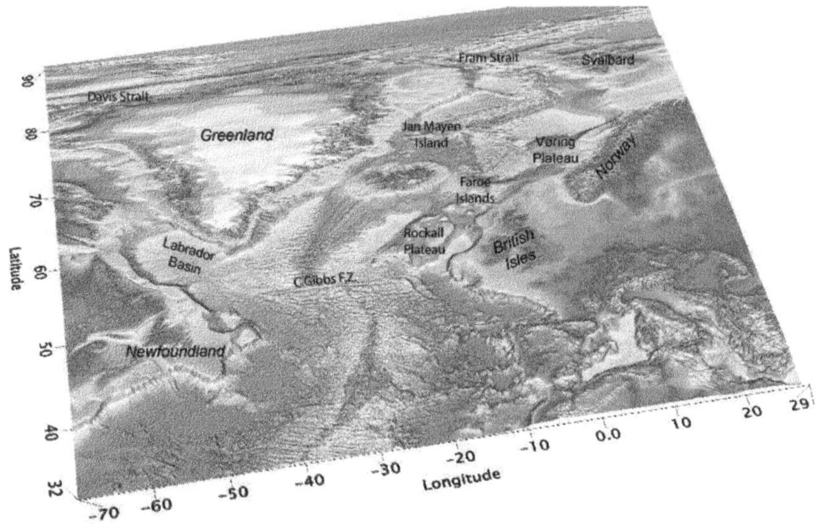

Figure 1.1. *The Northeast Atlantic Ocean and the Labrador–Baffin axis (Gebco 2019 base, L. Gernigon processing)*

These submerged domains are made up of lithospheres of oceanic and continental nature (section 3.2) with, sometimes, uncertain limits between these two types.

The emerged geology of Iceland shows a relatively recent differentiation. Large outcrops of so-called ancient basalts (15–5 My, in blue in Figure 1.2(b)) occupy coastal areas and are overlain by younger lavas (between 5 and 7 My, in green). This second generation of basalts is itself intersected by an active rift zone associated with Pleistocene (gray) and Holocene (pink) volcanism that crosses the whole island from south to north. In addition, four ice sheets cap presently the island in relation with especially wide brown bands (Figure 1.2(b)), which are the traces of subglacial volcanism, witnesses of a wider extension of the glaciers. It is also associated with a strong sedimentary splay along its southern coast (in light blue). Iceland has thus experienced a very complex history.

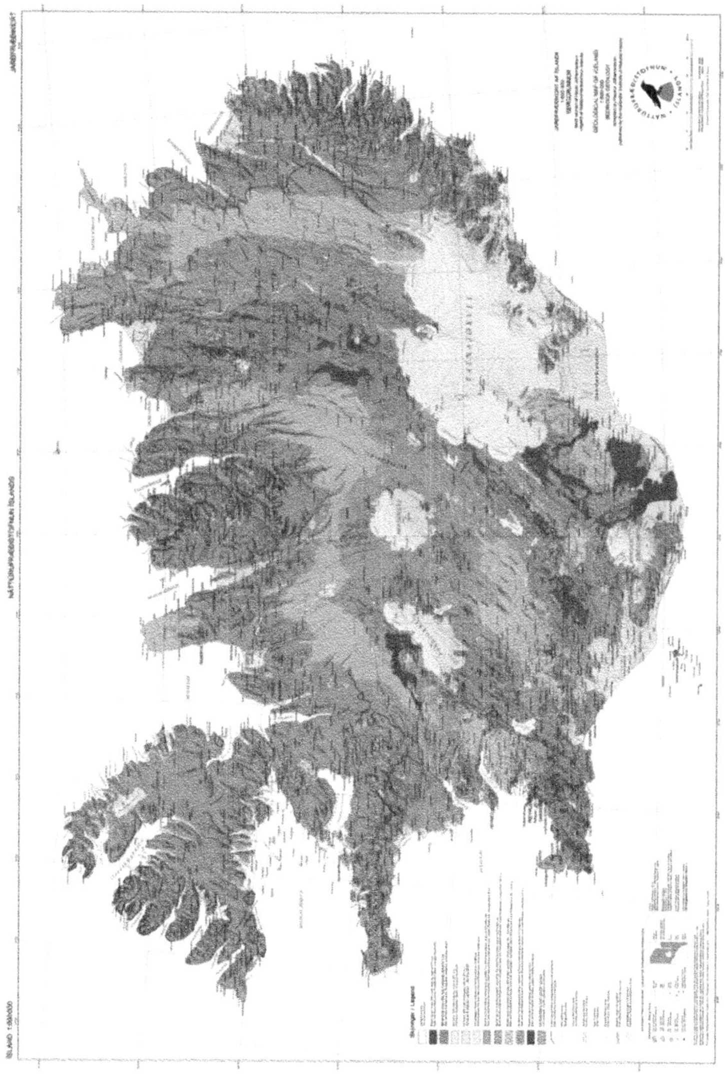

Figure 1.2. *(a) Geological map of Iceland, originally at 1/600,000e (H. Johanesson 2014). Náttúrufræðistofnun Ísland (Icelandic Institute of Natural History) (available at: https://en.ni.is/resources/publications/maps/geological-maps)*

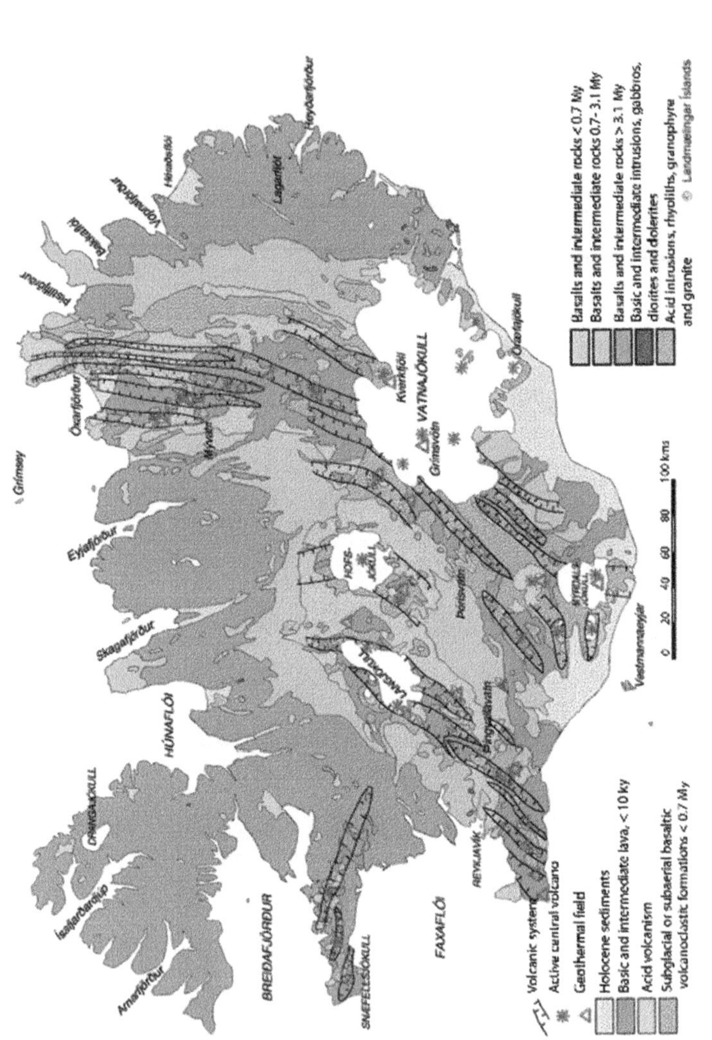

Figure 1.2. *(b) Simplified geological map of Iceland (Icelandic Geographical Institute/Landmælingar Ísland)*

1.1. Geographic and geodynamic context

The surface of Iceland appears as a vast plateau of about 600 m altitude (Figure 1.3, in orange), overlain by four large volcanic systems covered by four wide ice caps. Another surface, called strandflat, surrounds the island at about the level of the present coast and extends into the sea down to about –50 m; it is underlined by large swarms of small islands. Finally, a lowered submarine plateau surrounds the island to a depth of 300 m (Figure 1.3). The fact that it is cut by Neogene glacial scouring attests to its age (section 3.2 of Volume 2).

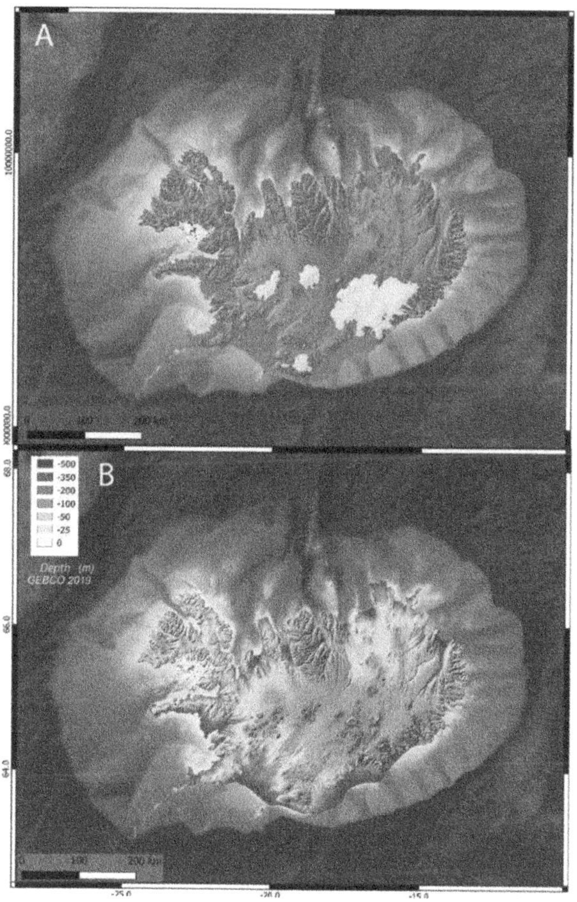

Figure 1.3. *Current morphology of Iceland. (a) A volcanic island shaped and occupied by glaciers and whose submarine plateau was indented by Neogene glaciations (GEBCO 2019 data). (b) Ice-free topography (modified from the "Digital Elevation Model" published by Bjorsson, 2017)*

The oceanic domains are fundamentally segmented: the basins are separated from each other by transform faults, such as the Ungava Fault in the Davis Strait between the Labrador Sea and Baffin Bay, or those of Charlie–Gibbs and Jan Mayen in the Northeast Atlantic, or by ridges transverse to the oceanic opening axes, such as the Greenland–Iceland–Faroe Islands Ridge. The entire region, including the Labrador-Baffin axis, Greenland and the Northeast Atlantic Ocean in the strict sense is referred to as the North Atlantic domain.

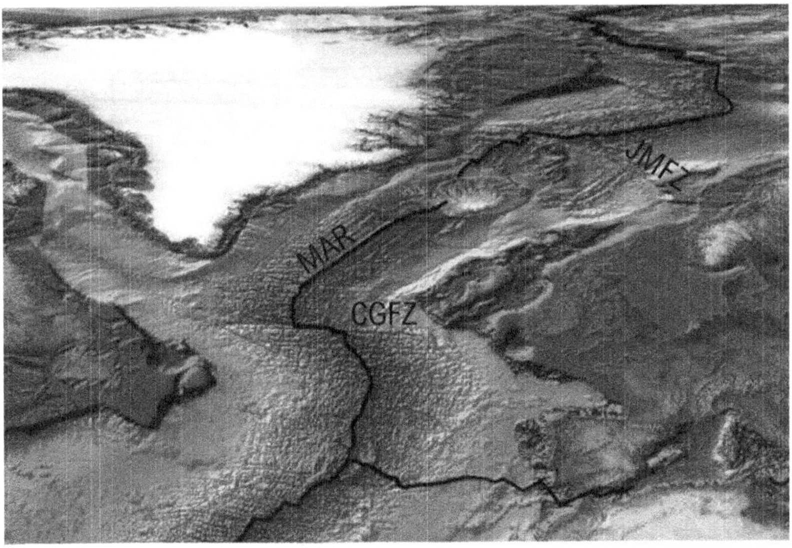

Figure 1.4. *The northern part of the Mid-Atlantic Ridge, modified from (Müller* et al. *2008)*

COMMENT ON FIGURE 1.4.– *The colors represent the age of the ocean floor, with red-orange corresponding to the period 20–0 million years ago. The opening of the North Atlantic began about 60 million years ago. CGFZ: Charlie–Gibbs Fracture Zone; JMFZ: Jan Mayen Fracture Zone; MAR: Mid-Atlantic Ridge.*

Iceland is generally presented as arising from the interaction between a thermal anomaly in the upper mantle, interpreted as a hot spot at the top of a plume, and a major axis of oceanic expansion, the Mid-Atlantic Ridge (MAR) (Figure 1.4).

The physical, chemical and dynamic characteristics of the Icelandic crust and lithosphere, and more generally of the Northeast Atlantic domain within which it is located, do not fit easily into simple models of continental break-up and accretion of the oceanic lithosphere. Understanding the origin of Iceland requires going back to the history of continental fragmentation between the North American and Eurasian plates in the Meso-Cenozoic (from 250 My). This history was characterized by the temporary individualization of a tectonic plate, Greenland, within a domain deeply marked by the heritage of the Caledonian collision (440–410 My).

1.2. Components of the North Atlantic domain

The major components of the North Atlantic domain are the MAR, the North Atlantic Igneous Province, the Icelandic hot spot and the GFIR.

1.2.1. *The Mid-Atlantic Ridge*

The MAR is a succession of ridge segments that range from the South Atlantic near Bouvet Island (latitude 54° S) to the North Atlantic, south of the Arctic Circle (latitude 87° N). In its southern part, it marks the boundary between the South American and African plates and in its northern part, it defines the boundary between the Eurasian and North American plates.

These segments are significantly offset by transform faults or zones. The two most important transform zones in the North-MAR are the Jan Mayen Fault Zone (latitude 71° N) and the Charlie–Gibbs Transform Zone (latitude 53° N), which limits to the south the North Atlantic oceanic domain *sensu stricto* (Figure 1.4). The Icelandic Rift, which represents the part of the ridge that emerged in Iceland, is itself currently shifted about 100 km eastward from the axis of the MAR (Chapter 2).

The North-MAR is a slow to ultra-slow ridge whose rate of expansion decreases toward the north (Le Breton *et al.* 2012). This rate varies from about 21 mm/year at the axis of the Reykjanes Ridge (slow ridge southwest of Iceland) to 16 mm/year at the Mohns Ridge (north of Jan Mayen). It becomes close to only 6 mm/year in the cold Arctic Ocean at the end of the ultra-slow Gakkel Ridge (Jokat *et al.* 2003), near the pole of rotation between the Eurasian and North American plates, located in the extreme East of Siberia. In Iceland, the rate of expansion at the axis of the ridge is of the order of 20 mm/year. The complex genesis of the North-MAR and its evolution during the Cenozoic are presented in Chapter 3 (section 3.2).

1.2.2. *The North Atlantic Igneous Province*

The North Atlantic Igneous Province (NAIP) is a basaltic province of the North Atlantic formed during the Paleogene. Before the opening of the North Atlantic, it extended over a large area (Saunders *et al.* 1997). Significant remnants of it exist mainly in Northern Ireland, Scotland, the Faroe Islands, and western Greenland (Figure 1.5 and Chapter 3).

At the end of the Cretaceous, the North America–Greenland–Europe region had no significant volcanic activity. Oceanic expansion was taking place between Canada and the southern Labrador Sea. Around 62 My, volcanic eruptions began in a vast region with extensive magma activity (intrusive and extrusive), notably in western Greenland (Chauvet *et al.* 2019) and in the sector of the Hebrides Islands (Wilkinson *et al.* 2016) and continued until 56 My.

1.2.3. *The Icelandic hot spot*

The concepts of hot spot and mantle plume, born at the same time as plate tectonics, are due to John Tuzo Wilson (1963) and Jason Morgan (1971). The first suggested, based on the case of Hawaiian volcanoes, that oceanic volcanic chains could constitute the trace left on a moving lithospheric plate of a fixed magma source located beneath this plate; the second proposed that the feeding of this fixed source was an abnormally hot plume rising from the base of the mantle (layer D''). Over the next half-century, geophysical and geochemical measurements, numerical modeling and laboratory experiments have refined and consolidated Wilson–Morgan's theory.

The link between plumes and the vast lava flows that constitute the large igneous provinces (traps and oceanic plateaus) was proposed in the 1980s (Courtillot *et al.* 1989), the idea being that, when the plumes come close to the surface, they thermally thin the lithosphere and generate very large magmatism. The volumes and rates of magmatic production are of the same order of magnitude as those of the ridges or arcs of the subduction zones. The NAIP would thus have been classically fed by a plume precursor to the one underlying the Icelandic hot spot.

Fluid mechanics have developed a "standard model" of a mantle plume, in which the plume is in the form of a long, narrow conduit – the tail of the plume – topped by a mushroom-shaped head below the lithosphere. Nevertheless, there are several variants of this model, some suggesting that the plumes flow toward and along the ridges through pipe-like channels, others through pancake-like gravity currents (see references to these works in Ito *et al.* 2003). The thickness of the oceanic crust in the Northeast Atlantic is abnormally high, as is the thickness of the mafic crust located between Greenland and the Faroe Islands. This is classically interpreted as a demonstration of abnormally high mantle melting rate, associated with a warmer than "normal" asthenospheric mantle (White and McKenzie 1989).

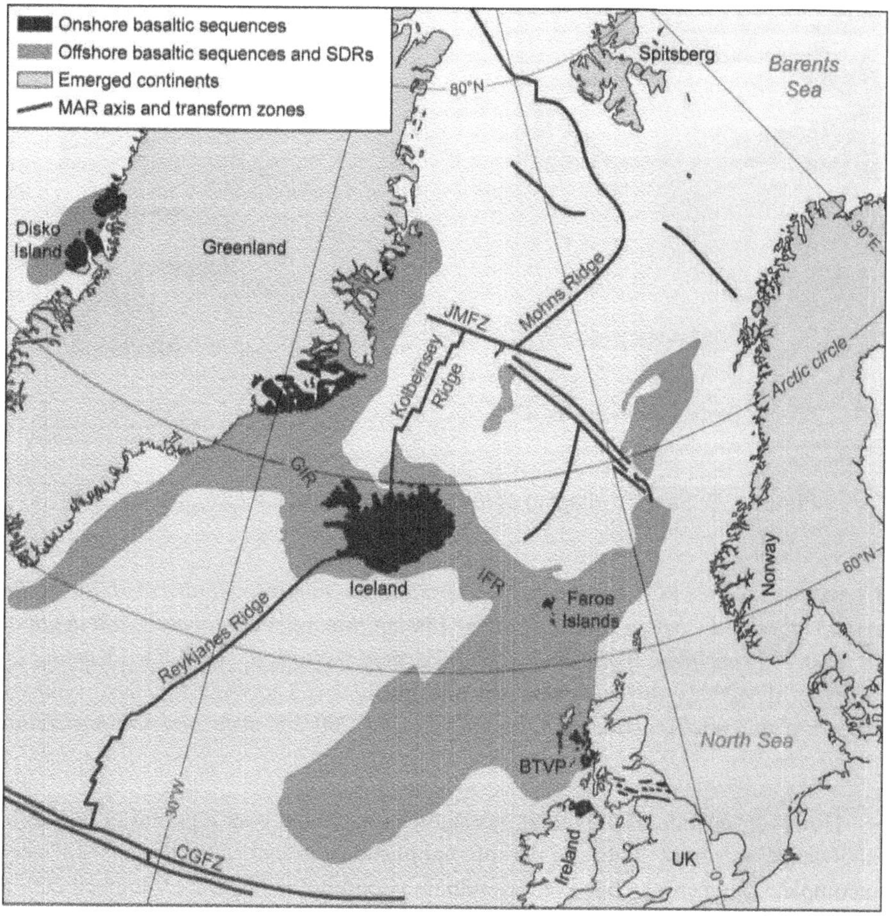

Figure 1.5. *The North Atlantic Igneous Province (from Storey* et al. *2007; Thodarson and Larsen 2007; Chauvet* et al. *2019)*

COMMENT ON FIGURE 1.5.– *GIR: Greenland–Iceland Ridge; IFR: Iceland–Faroe Islands Ridge, connecting the Icelandic hot spot with volcanic accumulations in the British Tertiary Volcanic Province (BTVP) south of the Faroe Islands; CGFZ: Charlie–Gibbs Fracture Zone; JMFZ: Jan Mayen Fracture Zone.*

Icelandic magmatism is commonly genetically related to the Paleogene magmatism of NAIP. It seems therefore related to a unique mantle plume that has existed for more than 60 My (Figure 1.6). Iceland could thus be interpreted as a residual plume axis (tail) whose "cap" is cooled.

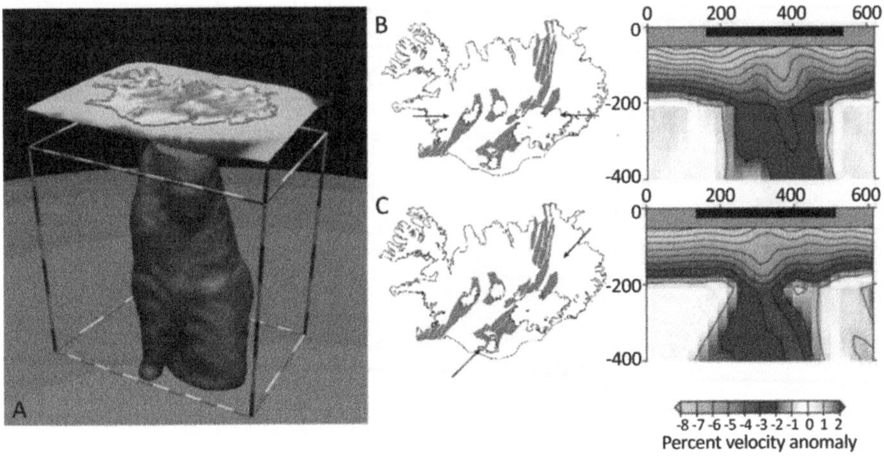

Figure 1.6. *Seismic imaging of the upper mantle at the Icelandic hot spot*

COMMENT ON FIGURE 1.6.– *Imagery illustrating the narrow cylindrical shape of the mantle plume (A), with a radius of about 150 km (low velocity anomaly to a depth of at least 400 km) (from Wolfe* et al. *1997). Vertical sections in the ICEMAN model by Allen* et al. *(2002) perpendicular (B) and parallel (C) to the rift. These sections illustrate the vertical conduit between 400 and 200 km deep and the horizontal plume above.*

However, in spite of the numerous studies carried out over a period of 50 years, knowledge about hot spots and the internal phenomena that cause them is still very incomplete. Numerous controversies remain (Courtillot *et al.* 2003; Campbell 2007). The spatial stability of hot spots is also questioned (Tarduno 2010) and the very reality of deep plumes is still debated (Anderson 2001; Montagner 2010).

As regards the Icelandic hot spot, no satisfactory model is currently available to account for all the very large amount of structural, tectonic, geochronological and petro-geochemical data available. The basic problem concerns the relationships between (i) a geochemical signature, suggesting an origin from the underlying deep mantle, (ii) the extensive deformation related to the ridges adjacent to and across the island, inducing adiabatic fusion at greater or lesser depths in the mantle, and finally (iii) the inheritance and recycling of ancient geological structures affecting the crust, such as the ancient oceanic suture and mantle in this particular area of the Atlantic, which may pollute the observed geochemical signatures.

The first oppositions to the deep plume model appeared in the early 2000s with the work of Foulger and her collaborators (Foulger *et al.* 2001, 2005; Foulger and Anderson 2005). For these authors, the large amount of magma produced in Iceland is thought to originate from the reworking of the subducted and eclogitized oceanic crust of the ancient Iapetus Ocean.

Several conferences, organized by pro- and/or anti-plume groups, were devoted to these issues during the first decade of the 21st century (in particular, a Penrose Conference held in Iceland in 2003, and the Great Plume Debate in Scotland in 2005), but the debate is far from over (Campbell and Kerr 2007; Meyer *et al.* 2007; Foulger 2012). Recent studies are still rehashing cards and, even among "pro-plume" authors, the spatial and temporal evolution of the latter is still the subject of work (Barnett-Moore *et al.* 2017; Martos *et al.* 2018). We will review the latest developments concerning the Icelandic hot spot in Chapter 3 and Chapter 1 of Volume 2.

1.2.4. *The Greenland–Iceland–Faroe Ridge*

The Greenland–Iceland–Faroe Ridge (GIFR) consists of a shoal interspersed with three fairly deep channels (about 2,000 m), which steers the North Atlantic from Greenland to the Faroe Islands, leaning against Iceland. It forms a relief of crucial importance not only as a recording of the last phases of the opening of the North Atlantic, but also for understanding the formation of the current thermohaline circulation, one of the major drivers of the evolution of our climate.

The GIFR crust has an anomalous thickness, ranging from 25 to 40 km, equivalent to that of a continental crust. It is not associated with any interpretable magnetic anomalies except locally between Iceland and the Faroes (Nuuns *et al.* 1983; Hjartarson *et al.* 2017).

1.3. Geodynamic characteristics of Iceland

One of the characteristics of Iceland is its sustained volcanic activity, accompanied by hydrothermal activity and almost permanent seismicity, although most of the time it is of low intensity and localized. Another major feature of the island lies in the presence of large glaciers and all that it implies from a geodynamic, climatic, touristic and industrial point of view.

1.3.1. *Seismicity*

While seismic activity in Iceland underlines all plate boundaries (Figure 1.7), it is less continuous and less strong in the rift and is mainly concentrated in the transform zones that connect the active rift to the offshore segments of the ridge to the north and south of the island.

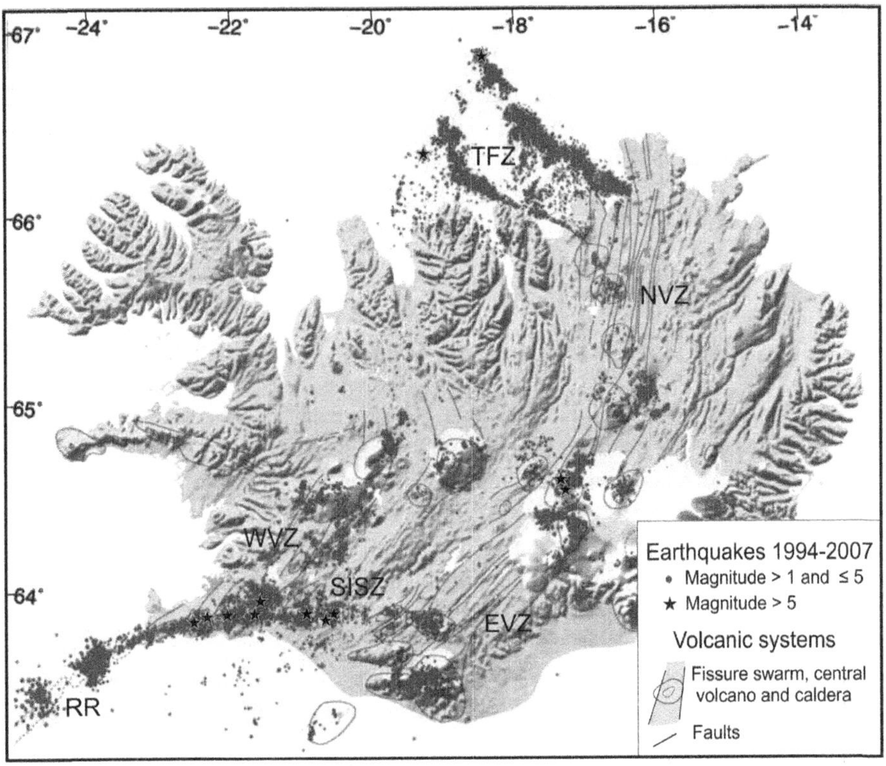

Figure 1.7. *Seismicity in Iceland recorded by the SIL network between 1994 and 2007 (modified from Jakobsdóttir 2008). Volcanic systems according to Einarsson and Sæmundsson (1987). EVZ: East Volcanic Zone; NVZ: North Volcanic Zone; WVZ: West Volcanic Zone; ZFT: Tjörnes Fracture Zone; SISZ: South Iceland Seismic Zone*

The Icelandic Rift corresponds to the onshore extensions of two segments of the North-MAR, the Reykjanes Ridge to the south and the Kolbeinsey Ridge to the north (Figures 1.2 and 1.5). It has three branches (Figure 1.7), all offset eastward from the axis of the North-MAR. These branches of the rift consist of fissure swarms associated with active central volcanoes, a significant portion of which are located beneath the major glaciers of the island; they can be guessed from the ice-free topographic map (Figure 1.2(b)). In the rift, seismicity is strongly linked to the functioning of magmatic systems (Figure 1.7). This seismicity of volcanic origin and the volcanism itself (for example, the recent eruption of the Bárðarbunga volcano and its large flow of basaltic lava) testify to an important active magmatism supposed to be fed by the Icelandic hot spot.

The shift of the Icelandic rift with respect to the axis of the North-MAR is accommodated by two transform zones, the South Iceland Seismic Zone (SISZ) to the south and the Tjörnes Fracture Zone (TFZ) to the north. Both are inherited from an older history, that of the opening of the Atlantic (Chapter 3) and linked to the "rift jump" phenomenon (Chapter 2). It is in these two transform zones that seismicity is the most sustained (Figure 1.7) with both quasi-permanent microseismicity and major recurrent seismicity whose magnitude can exceed 7 (section 2.2.1).

The disposition of the two rift branches in the south of the island (WVZ and EVZ) resembles that of the overlapping spreading centers (OSC) described on fast-spreading ridges, rather than that of a true transform zone. Nevertheless, the associated seismicity characterizes well the left-lateral transform motion of the SISZ, connecting the EVZ to the Reykjanes Ridge. At present, the two branches of the rift do not have the same level of activity, and the WVZ accommodates only a small part of the extension; the question of whether this present low activity is representative of the last thousands or millions of years will be discussed in Chapter 2.

1.3.2. *Icelandic volcanism*

Since its insularization, 16 My ago, Iceland has formed a volcanic land located in a central position in the Atlantic Ocean, a site subject to numerous effusive fissural eruptions. Therefore, it is also one of the main areas of the northern hemisphere generating explosive eruptions emitting into the atmosphere ashes, tephra and gases.

Icelandic quaternary volcanism allows us to understand, by its outcrop along coastal cliffs or canyons, the mechanisms of formation of numerous eruptive figures in aerial or underwater basaltic environments. This is the case of rootless volcanoes and lava lakes (section 1.2 of Volume 2), the establishment of dykes or sills and their control by tectonics, or the evolution of stresses related to glaciations. We have shown that close mechanical and physical links probably exist between deglaciation, volcanism, seismicity and tectonic deformations (section 2.1 of Volume 2). The key process lies in variations in the adiabatic melting rate of the mantle near the lithosphere/asthenosphere boundary, induced by decompression related to crustal extension or glacio-isostatic discharge. The emission zones of these lava flows have therefore constrained the evolution of Iceland since the onset of glaciations. It also enabled the understanding of the formation of SDRs (Seaward Deeping Reflectors), a mode of accumulation of lava flows especially recognized by seismic reflection along the margins of the North Atlantic.

Tephra successions are commonly recorded in the cores drilled in the Greenland ice cap and in marine sediments deposited on the bottom of the North Atlantic. Some of these tephras were distally dated by radiocarbon as in the peat bogs of northern Europe, others by age patterns in ice or marine sedimentary accumulations (sections 1.3 and 3.2 of Volume 2). More rarely, their dating has been performed directly by K-Ar analysis (section 2.4 of Volume 2). These emissions of tephra and associated gases have a significant environmental impact on ocean fertility, as the GIFR seabed is an important breeding ground for the majority of fish consumed, as well as on the health of human or animal populations and on air safety, as shown by the recent eruption of Eyjafjallajökull (2010).

1.3.3. *Eustatism and the Icelandic glaciers*

The full insularization of Iceland began about 16 My ago (Figures 1.2 and 1.8). The evolution of Iceland until around 9 My was mainly controlled by the tectonic and magmatic processes related to both the opening of the Atlantic Ocean and the evolution of the hot spot. They built a classically evolving volcanic island, even if its formation was unusual (Chapter 3). The last rift jump (Chapter 2) in the north of the island started about 8 My ago. Then a rift jump occurred in the south of the island around 3 My, when the great glaciations started. At the same time, Iceland's shape was also influenced by global eustatic evolution, which allowed the development of marine abrasion surfaces around the island (Figures 1.8 and 1.9).

The development of radiometric dating (section 2.4 of Volume 2) allowed us to better constrain both the chronology of the progressive extinction of the paleorifts and also that of the glaciations with respect to the tectonic and paleo-oceanographic evolution of the North Atlantic. The impact of the glacial loading of rifts during periods of extended ice caps could explain *pro parte* this cyclic evolution linked to rift-jump.

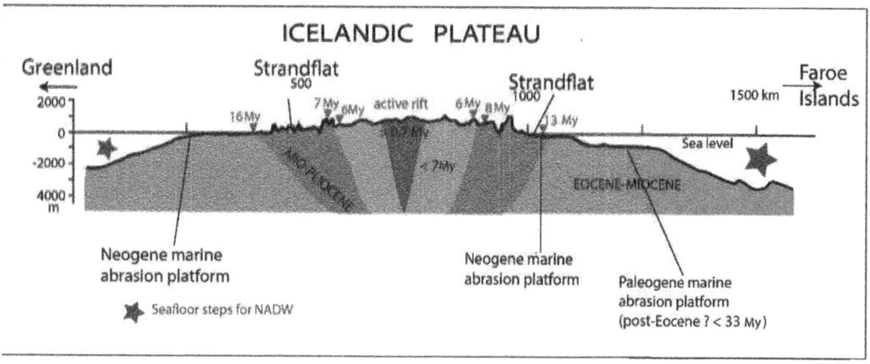

Figure 1.8. *W-E topographic section of Iceland with its various morphological and geological dated constituents. The geology is partially hypothetical, adapted and completed from (Hjartason* et al. *2007). K-Ar dates are in red. NADW: North Atlantic Deep Water*

The upper plateau of the island, raised at about 600 m in altitude, is a relatively recent construction (<7 My) as far as the age of dyke injection and subaerialvolcanism is concerned. The peninsulas of Flateyjarskagi, Tröllaskagi and Tjörnes are mainly constituted by lava flows dated between 5 and 3 My, interspersed with glacial and marine formations, as in the sector of Husavik. The peninsulas of the northwest and east consist of much older lavas. The ages of these basaltic series usually range between 7 and 6 My very locally reaching 16 My at their base (in the west).

Strandflat is a marine abrasion surface that can be found along the whole North Atlantic. It has sometimes been attributed to the action of glaciers only, but is scoured both on the edge of continents and in Iceland by large glacial valleys that reuse a pre-existing hydrographic network. Between +20 and –50 m, the platform of the Icelandic strandflat contains mostly lava flows dated between 7 and 10 My. It is strongly shaped by the action of glaciers. Several ages, obtained from interglacial flows fossilizing the strandflat, are distributed between 1.8 My on Skagi to 0.13 My for Snæfellnes: it is therefore polygenetic. It is also continuous over several dozen square kilometers between the three great peninsulas of the west coast. In the rest of the island, it forms a relatively narrow terrace.

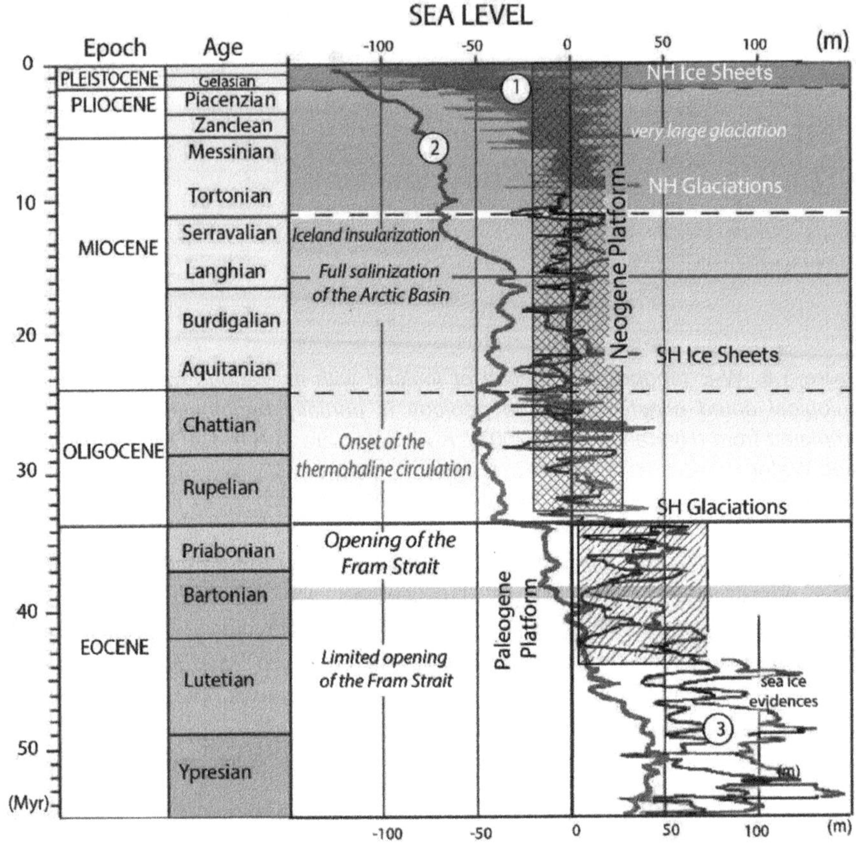

Figure 1.9. *Global sea level evolution from the Eocene to the present day in relation to glaciations and the development of thermohaline circulation*

COMMENT ON FIGURE 1.9.– *(1) Compilation of sea level changes (based on (Abreu* et al. *1998)). (2) Curve* $\delta^{18}O$ *compiled by Cramer* et al. *(2008). (3) Eustatic curves defined by Miller* et al. *(2005, 2011, in black and blue) and Kominz* et al. *(2008, in brown). The blue background reflects the intensification of the thermohaline circulation. SH: Southern hemisphere, NH: Northern hemisphere. It can be compared with Figures 3.6. and 3.3(B) in Volume 2.*

The submerged platform that lies between –200 m and –50 m around the island (Figures 1.1 and 1.3) truncates the Miocene and Oligocene effusive formations of the GIR (Hjartason *et al.* 2007; Chapter 3 of this volume). This platform is itself scoured by glaciers (Figures 1.3 and 1.10) as is the West Coast of Spitsbergen or the coast of Newfoundland (Figure 3.1(A)), which are free of volcanic activity.

The evolution of the eustatic signature (Miller *et al.* 2005, 2011; Kominz *et al.* 2008) on long time scales indicates that the tectono-eustatic component has not exceeded 50 m since 60 My (Rowley 2013), mainly as a result of variations in seafloor production and mantle evolution.

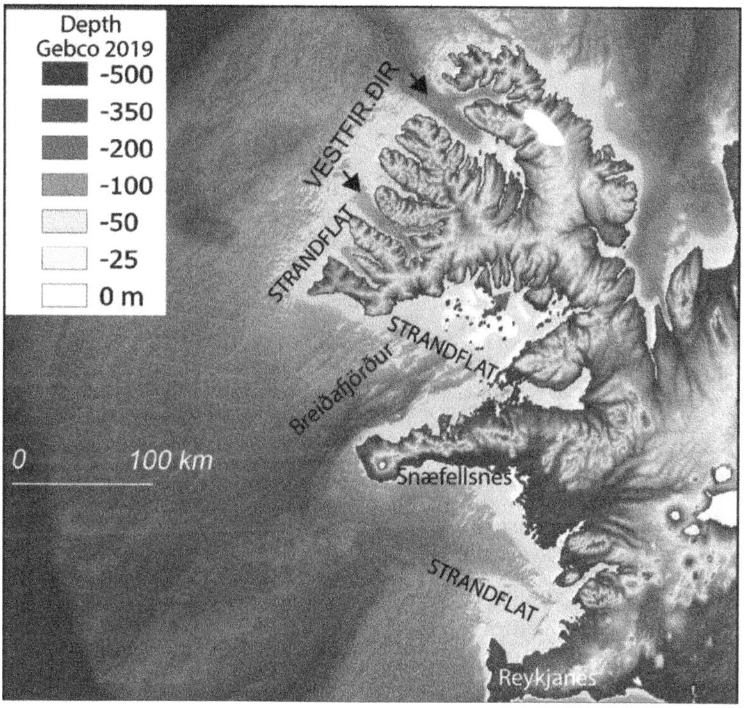

Figure 1.10. *The Northwest Peninsula, western Iceland, and the strandflat*

COMMENT ON FIGURE 1.10.– *Subsident morphology with a very large strandflat (<50 m, clear) south of the Northwest Peninsula and west of Iceland with fossilized coastal cliffs, flooded glacial valleys up to the upstream cirque (red arrow) and eroded platform in the Vestfirðir sector. Neogene and Quaternary glacial overdeepenings are indicated by black arrows.*

In fact, this surface could have been shaped in approximately the Lower Oligocene and especially at the end of this epoch (Chattian). The first evidence of the glaciations of the northern hemisphere dates back to 9 My (beginning of the Tortonian) thus trace since this period a thermal subsidence and a tilting of the platform of at least 150 m west of Iceland, allowing an accentuation of the thermohaline circulation (section 3.1 of Volume 2). The flooded glacial valleys of the southern part of the Northwest Peninsula (Figure 1.10) also illustrate this evolution.

The eastern coast of the island, on the other hand, has a different history. It includes an older surface, located today at a depth of 1,000 m, which may have been shaped in the Bartonian age, a key period for the tectonics of the Atlantic margins (Chapter 3).

1.4. References

Allen, R.M., Nolet, G., Morgan, W.J., Vogfjörd, K., Bergsson, B.H., Erlendsson, P., Foulger, G.R., Jakobsdóttir, S.S., Julian, B.R., Pritchard, M., Ragnarsson, S., Stefánsson, R. (2002). Imaging the mantle beneath Iceland using integrated seismological techniques. *J. Geophys. Res. Atmos.*, 107(12) [Online]. Available at: http://doi.org/10.1029/2001JB000595.

Anderson, D.L. (2001). Top-down tectonics? *Science*, 293, 2016–2018.

Andresen, A., Rehnström, E.F., Holte, M.K. (2007). Evidence for simultaneous contraction and extension at different crustal levels during the Caledonian orogeny in NE Greenland. *J. Geol. Soc. London*, 164, 869–880.

Barnett-Moore, N., Hassan, R., Müller, R., Williams, S.E., Flament, N. (2017). Dynamic topography and eustasy controlled the paleogeo-graphic evolution of northern Africa since the mid-Cretaceous. *Tectonics*, 36, 929–944.

Bessin, P., Guillocheau, F., Robin, C., Schroëtter, J.M., Bauer, H. (2015). Planation surfaces of the Armorican Massif (Western France): Denudation chronology of a Mesozoic land surface twice exhumed in response to relative crustal movements between Iberia and Eurasia. *Geomorphology*, 233, 75–91.

Campbell, I.H. (2007). Testing the plume theory. *Chem. Geol.*, 241(3–4), 153–176.

Campbell, I.H. and Kerr, A.C. (2007). The Great Plume Debate: Testing the plume theory. *Chem. Geol.*, 241(3–4), 149–152.

Chauvet, F., Geoffroy, L., Guillou, H., Maury, R.C., Le Gall, B., Agranier, A., Aviana, A. (2019). Eocene continental breakup in Baffin Bay. *Tectonophysics*, 757, 170–186.

Courtillot, V., Richards, M., Duncan, R. (1989). Flood basalts and hotspot tracks: Plume heads and tails. *Science*, 246(4926), 103–107.

Courtillot, V., Davaille, A., Besse, J., Stock, J. (2003). Three distinct types of hotspots in the Earth's mantle. *Earth Planet Sci. Lett.*, 205, 295–308.

Cramer, B.S., Toggweiler, J.R., Wright, J.D., Katz, M.E., Miller, K.G. (2009). Ocean overturning since the late Cretaceous: Inferences from a new benthic foraminiferal isotope compilation. *Paleoceanography*, 24(14) [Online]. Available at: http://doi.org/10.1029/2008PA001683.

Einarsson, P. and Sæmundsson, K. (1987). Earthquake epicenters 1982–1985 and volcanic systems in Iceland. In *Í hlutarins eðli. Festschrift for Þorbjörn Sigurgeirsson*, Sigfússon, Þ.I. (ed.). Menningarsjóður, Reykjavík.

Foulger, G.R. (2012). Are "hot spots" hot spots? *J. Geodin.*, 58, 1–28.

Foulger, G.R. and Anderson, D.L. (2005). A cool model for the Iceland hotspot. *J. Volcan. Geotherm. Res.*, 141, 1–22.

Foulger, G.R., Pritchard, M.J., Julian, B.R., Evans, J.R., Allen, R.M., Nolet, G., Morgan, W.J., Bergsson, B.H., Erlendsson, P., Jakobsdóttir, S., Ragnarsson, S., Stefánsson, R., Vogfjörð, K.S. (2001). Seismic tomography shows that upwelling beneath Iceland is confined to the upper mantle. *Geophys. J. Inter.*, 146, 504–530.

Foulger, G.R., Natland, J.H., Anderson, D.L. (2005). A source for Icelandic magmas in remelted Iapetus crust. *J. Volcan. Geotherm. Res.*, 141, 23–44.

Hjartarson, Á., Erlendsson, Ö., Blischke, A. (2017). The Greenland–Iceland–Faroe Ridge complex. In *The NE Atlantic Region: A Reappraisal of Crustal Structure, Tectonostratigraphy and Magmatic Evolution*, Péron-Pinvidic, G., Hopper, J.R., Stoker, M., Gaina, C., Funck, T., Árting, U.E., Doornenbal, J.C. (eds). Geological Society of London, London.

Jakobsdóttir, S.S. (2008). Seismicity in Iceland: 1994–2007. *Jökull*, 58, 75–100.

Jokat, W., Ritzmann, O., Schmidt-Aursch, M.C., Drashev, G.S., Snow, J.E. (2003). Geophysical evidence for reduced melt production on the Arctic ultraslow Gakkel mid-ocean ridge. *Nature*, 423(6943), 962–965.

Lisiecki, L.E. and Raymo, M.E. (2005). A Pliocene-Pleistocene stack of 57 globally distributed benthic δ18O records. *Paleoceanography*, 20 [Online]. Available at: http://doi.org/10.1029/2004PA001071.

Kominz, M.A., Browning, J.V., Miller, K.G., Sugarman, P.J., Mizintseva, S., Scotese, C.R. (2008). Late Cretaceous to Miocene sea-level estimates from the New Jersey and Delaware coastal plain coreholes: An error analysis. *Basin Res.*, 20(2), 211–226.

Martos, Y.M., Catalan, M., Jordan, T.A., Golynsky, A., Golynsky, D., Eagles, G., Vaughan, D.G. (2017). Heat flux distribution of Antarctica unveiled. *Geophys. Res. Lett.*, 44, 11417–11426.

Meyer, R., Van Wijk, J., Gernigon, L. (2007). The North Atlantic Igneous Province: A review of models for its formation. *Geol. Soc. Amer.*, 430, 525–552.

Miller, K.G., Kominz, M.A., Browning, J.V., Wright, J.D., Mountain, G.S., Katz, M.E., Sugarman, P.J., Cramer, B.S., Christie-Blick, N., Pekar, S.F. (2005). The Phanerozoic record of global sea-level change. *Science*, 310, 1293–1296.

Miller, K.G., Mountain, G.S., Wright, J.D., Browning, J.V. (2011). A 180-million-year record of sea level and ice volume variations from continental margin and deep-sea isotopic records. *Oceanography*, 24(2), 40–53.

Montagner, J.P. (2010). Panaches chauds : mythe ou réalité ? *Pour la science : la terre à cœur ouvert*, 67, 46–54.

Morgan, W.J. (1971). Convective plumes in the lower mantle. *Nature*, 230, 42–43.

Nunns, A.G., Talwani, M., Lorentzen, G.R., Vogt, P.R., Sigurgeirsson, T., Kristjánsson, L., Larsen, H.C., Vopel, D. (1983). Magnetic anomalies over Iceland and surrounding seas. In *Structure and Development of the Greenland–Scotland Ridge*, Bott, M.H.P., Saxov, S., Talwani, M., Thiede, J. (eds). Plenum, New York.

Rehnström, E.F. (2010). Prolonged Paleozoic magmatism in the East Greenland Caledonides: Some constraints from U-Pb ages and Hf isotopes. *Journal of Geology*, 118, 447–465.

Rowley, D.B. (2013). Sea level: Earth's dominant elevation. Implications for duration and magnitudes of sea level variations. *J. Geol.*, 21(5), 445–454.

Saunders, A.D., Fitton, J.G., Kerr, A.C., Norry, M.J., Kent, R.W. (1997). The North Atlantic Igneous Province. In *Large Igneous Provinces: Continental, Oceanic, and Planetary Flood Volcanism*, Mahoney, J.J. and Coffin, M.F. (eds). American Geophysical Union, Washington.

Tarduno, J. (2010). La mobilité des points chauds. *Pour la science : la terre à cœur ouvert*, 67, 56–61.

Van Vliet-Lanoë, B., Vandenberghe, N., Laignel, B., Laurent, M., Lauriat-Rage, A., Louwye, S., Mansy, J.L., Meilliez, F., Mercier, D., Hallegouët, B., Laquement, F., Michel, Y., Moguedet, G. (2002). Paleogeographic evolution of the Western Europe during the upper Cenozoic. *Geodiversitas*, 24(3), 511–541.

White, R.S. and Mackenzie, D. (1989). Magmatism at rift zones: The generation of volcanic continental margins and flood basalts. *J. Geophys. Res.*, 94, 7685–7729.

White, R.S., Spence, G.D., Fowler, S.R., McKenzie, D.P., Westbrook, G.K., Bowen, A.N. (1987). Magmatism at rifted continental margins. *Nature*, 330, 439–444.

Wilkinson, C.M., Ganerød, M., Hendriks, B.H.K., Eide, E.A. (2016). Compilation and appraisal of geochronological data from the North-Atlantic Igneous Province (NAIP). In *The NE Atlantic Region: A Reappraisal of Crustal Structure, Tectonostratigraphy and Magmatic Evolution*, Péron-Pinvidic, G., Hopper, J.R., Stoker, M., Gaina, C., Funck, T., Árting, U.E., Doornenbal, J.C. (eds.). Geological Society of London, London [Online]. Available at: http://doi.org/10.1144/SP447.10.

Wilson, J.T. (1963). A possible origin of the Hawaiian Islands. *Can. J. Physics*, 41, 863–870.

Wolfe, C.J., Bjarnason, I.T., Van Decar, J.C., Solomon, S.C. (1997). Seismic structure of the Iceland mantle plume. *Nature*, 385, 245–247.

2

Iceland, an Emerging Ocean Rift

Françoise BERGERAT with the collaboration of Brigitte VAN VLIET-LANOË

2.1. Mid-Atlantic Ridge and Icelandic hot spot interactions

Regardless of the nature and origin of hot spots – and in particular of the Icelandic hot spot – and whether they are fixed or mobile, the fact remains that there is a major thermal anomaly under Iceland, commonly referred to as the "Icelandic hot spot". The overlying lithospheric plates are moving relative to it and – if it is assumed that it is mobile – at a different (faster) speed. When an oceanic ridge is located above such a hot spot, as is the case in Iceland, overabundant volcanism will occur along the ridge, a feature that accounts for the considerable lava effusions, the abnormally thick oceanic crust (20–30 km) and the emergence of the island about 16 My ago. In the theoretical case shown in Figure 2.1, the ridge migrates increasingly further away from the hot spot, and volcanism continues to occur there; the volcanic alignments left in the lithospheric plate thus represent the traces left by the mantle plume (Figures 2.1(A), (B1) and (C)). What happened in Iceland is somewhat different from this process. The Mid-Atlantic Ridge migrated westward relative to the Icelandic hotspot, but after a few million years, the westward-trailing part of the rift gradually became inactive, and a new rift was created above the apex of the mantle plume (Figure 2.1(B2)). This phenomenon is called rift jump (Ward 1971; Sæmundsson 1974).

For color versions of the figures in this chapter see www.iste.co.uk/vanvliet/iceland1.zip.

The capture mechanism of the rift by the hot spot as well as the rift jump process are still very poorly known, and the ancient positions of the rift are themselves still much discussed (section 2.3.1.1). The existence of paleo-rifts is classically admitted in the Húnaflói-Skagi and Snæfells peninsulas (Sæmundsson 1974) (Figure 2.2).

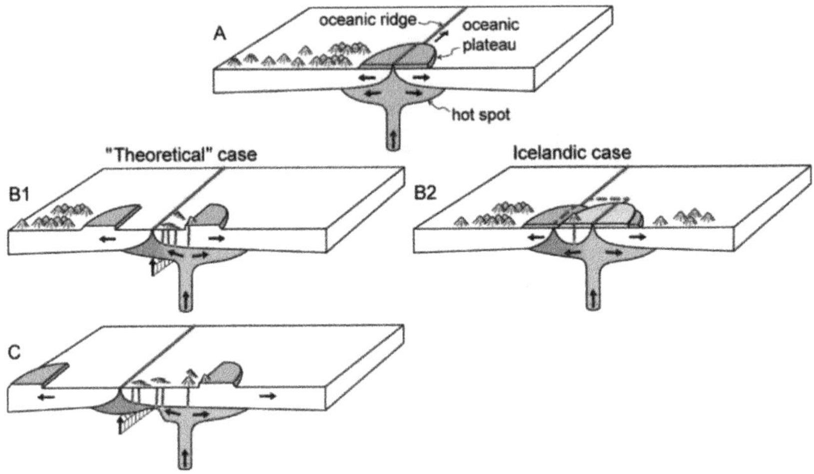

Figure 2.1. *Interactions between an oceanic ridge and a hot spot, and the specific case of Iceland (A, B1 and C modified from Dyment* et al. *2007)*

COMMENT ON FIGURE 2.1.– *We consider a ridge initially located above a hot spot, with the creation of an oceanic plateau (A). In the "theoretical" case, the ridge gradually moves away from the hot spot while continuing to function (B1), then finally it completely escapes the influence of the latter (C). In the case of the Mid-Atlantic Ridge and the Icelandic hot spot, when the ridge moves away from the hot spot, it gradually ceases to function at the latitude of the island (red dashed line), while a new rift is created above the hot spot. A transform zone is formed to connect the new rift to the ridge (red dashed line) (B2).*

The Icelandic rift is, today, located more than 100 km away from the accretion axis of the oceanic ridge. It connects to the northern and southern segments of the ridge through two transform zones (Figure 2.2), the Tjörnes Fracture Zone (TFZ) and the South Iceland Seismic Zone (SISZ). These zones possibly reactivate zones initiated earlier, since the formation, at the beginning of the Eocene, of the Reykjanes Ridge (RR) and, at the beginning of the Oligocene, of the Kolbeinsey Ridge (KR) connected to the Greenland-Faroe Ridge (see Chapter 3 as well as Le Breton *et al.* 2012; Hjartarson *et al.* 2017).

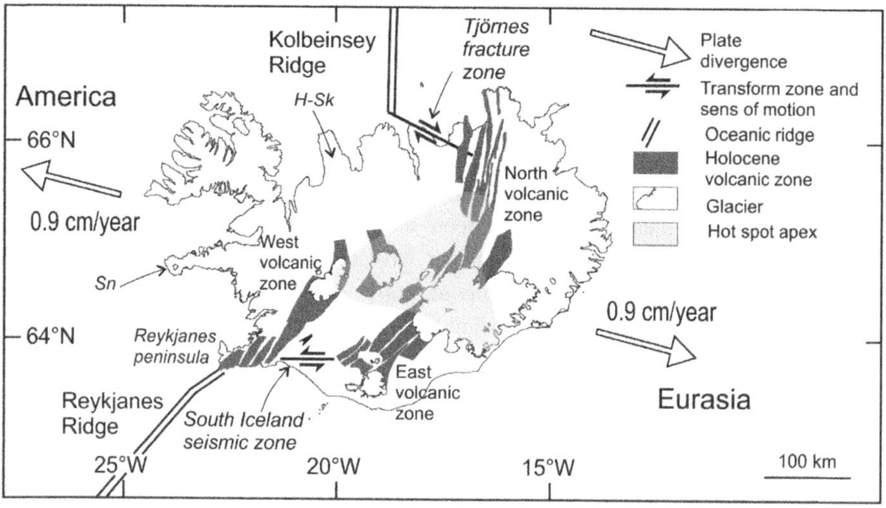

Figure 2.2. *Plate boundaries in Iceland and major structural zones. Holocene volcanic systems according to Sæmundsson (1979). Plate divergence from Nuvel-1 data (DeMets* et al. *1990, 1994). Hot spot trace at 400 km depth according to (Tryggvason* et al. *1983). H-Sk: Húnaflói-Skagi peninsula; Sn: Snæfells peninsula*

The active rift is composed of three major volcano-tectonic zones (Figure 2.2). The two most active, the North Volcanic Zone (NVZ) and the East Volcanic Zone (EVZ) are centered above the hot spot apex, while the third one, the West Volcanic Zone (WVZ), is considered as a portion of the rift in the process of extinction. Since 1–2 My, the NVZ has propagated northwards and the KR southwards (e.g. Guðmundsson 2007). Similarly, the southern part of the EVZ is commonly considered to be a southward spreading part of the rift since about 3 My.

2.2. Present-day deformations in Iceland

Present-day deformations in Iceland are well documented by data acquired through the networks established and developed since the late 1980s: (i) the network of permanent seismic stations installed by the IMO (Icelandic Meteorological Office or Veðurstofa Íslands) and providing a large numerical database (e.g. Stefánsson *et al.* 1993) and (ii) a CGPS (Communicative Global Positioning System) network, which is gradually being deployed (Geirsson *et al.* 2006) and is the result of collaboration between various Icelandic, American and European institutes and universities (Háskóli Íslands, IMO, IGS, Landmaelingar Íslands, NordVulk, Université de Savoie, University of Arizona). This network currently comprises more than 90 stations and is operated by the IMO. National GPS campaigns (ÍSNET) also took place in 1993 and 2004 with the measurement of 120 sites evenly distributed throughout the island (Árnadóttir *et al.* 2006). Finally, InSAR (Interferometric Synthetic Aperture Radar images) interferometric analysis is as well widely used. All these data have been extensively used to analyze cosismic deformations, ground deformations, glacier isostatic adjustments and plate motions (Árnadóttir *et al.* 2001, 2004; Pedersen *et al.* 2001, 2003; Pagli *et al.* 2003; La Femina *et al.* 2005; Geirsson *et al.* 2006; Jouanne *et al.* 2006; Jónsson 2008; Sudhaus and Jónsson 2008; Hreinsdóttir *et al.* 2014; Reverso *et al.* 2014).

2.2.1. *Seismicity*

Seismicity in Iceland occurs mostly along the plate boundaries, i.e. along the rift zones, where it is strongly linked to volcanism, and mainly in the two transform zones. It is indeed in the TFZ and SISZ that the seismicity is the most significant both in terms of the number of earthquakes and their intensity (Figure 1.6). There, the occurrence of a large number of major earthquakes is coupled with permanent microseismicity (Stefánsson *et al.* 1993; Jakobsdóttir *et al.* 2002; Jakobsdóttir 2008).

This section concerns the very recent to present period, since the first instrumental seismic records on the European continent, i.e. since the end of the 19th century. Historical and paleo-earthquakes, whose study is based on field observations rather than on instrumentation, will be detailed in section 2.3.2.

Much of the information concerning the earthquakes that occurred between 1950 and the installation of the Icelandic network come from the United States Geological Survey (USGS). Seismic activity in Iceland has been regularly recorded for about 30 years by the SIL seismic network, managed by IMO (Stefánsson *et al.* 1993), which currently consists of 42 digital seismic stations coupled with a data processing system (Jakobsdóttir *et al.* 2002). It has been progressively installed – and its detection capability improved – since the early 1990s along active seismic zones (transform zones) and rift zones (Þorbjarnardóttir and Guðmundsson 2003). It records earthquakes down to a magnitude[1] below zero. Since earthquakes of magnitude greater than 4 tend to be locally underestimated, these magnitudes are

1 An earthquake's magnitude is a logarithmic representation of its "size", proportional to the seismic energy released. The magnitudes indicated in this book are those given by the cited authors (M, M_W, M_L, M_b or M_S).

also calculated by the National Earthquake Information Center (NEIC) in the United States. In general, the depth of earthquakes in Iceland is less than 8–12 km, which corresponds to the thickness of the brittle crust. Deeper earthquakes are sometimes recorded (up to 25–30 km), but they are poorly constrained, their epicenters being located far from the stations, and the velocity models are not detailed enough (Jakobsdóttir 2008).

2.2.1.1. *Seismicity of volcanic zones*

Seismicity in volcanic zones is characterized by fairly constant activity. Magma intrusions induce swarms of occasional low magnitude earthquakes (Pedersen *et al.* 2007; Jakobsdóttir *et al.* 2008; Soosalu *et al.* 2010; Plateaux *et al.* 2012). Most of the seismicity is at a depth of around 10 km, but earthquakes sometimes occur between 15 and 20 km, close to central volcanoes. The magnitudes of these earthquakes are generally small ($0 < M < 2$), and they are probably associated with processes occurring in deep magmatic systems or within the roots of volcanoes (Soosalu *et al.* 2010; Greenfield *et al.* 2016).

Historical records show that the most active volcanoes in Iceland (80% of eruptions) are Hekla, Grímsvötn, Katla and Bárðarbunga (Þórðarson and Larsen 2007; Þórarinsson and Sæmundsson 1979) (see location in Figure 2.14).

In recent years, Katla has been extremely seismic, with dozens of earthquakes recorded every week. Its seismicity is generally shallow and appears to be related to geothermal processes, surface intrusions, glacier deformation and landslides (Einarsson and Brandsdóttir 2000; Jónsdóttir *et al.* 2007, 2009; Sgattoni *et al.* 2017).

Hekla is seismically quite calm, as only a few earthquakes are recorded each month, and even the seismicity that preceded the 1991 and 2000 eruptions remained limited (Ágústsson *et al.* 2000; Soosalu *et al.* 2005; Sturkell *et al.* 2006). In contrast, Grímsvötn's seismicity was significant in the months preceding the 2004 and 2011 eruptions (Vogfjörð *et al.* 2005).

Good quality seismic data are difficult to obtain for the volcanoes located beneath Vatnajökull due to the lack of bedrock for installing the stations and the lack of stability of electrical or telecommunications installations. Earthquakes of small magnitude are not recorded, and location errors can be significant. Despite this, the earthquakes location shows that seismic activity is mainly concentrated on the central volcanoes: Grímsvötn, Öræfajökull, Bárðarbunga, Kverkfjöll, Esjufjöll and Þórðarhyrna.

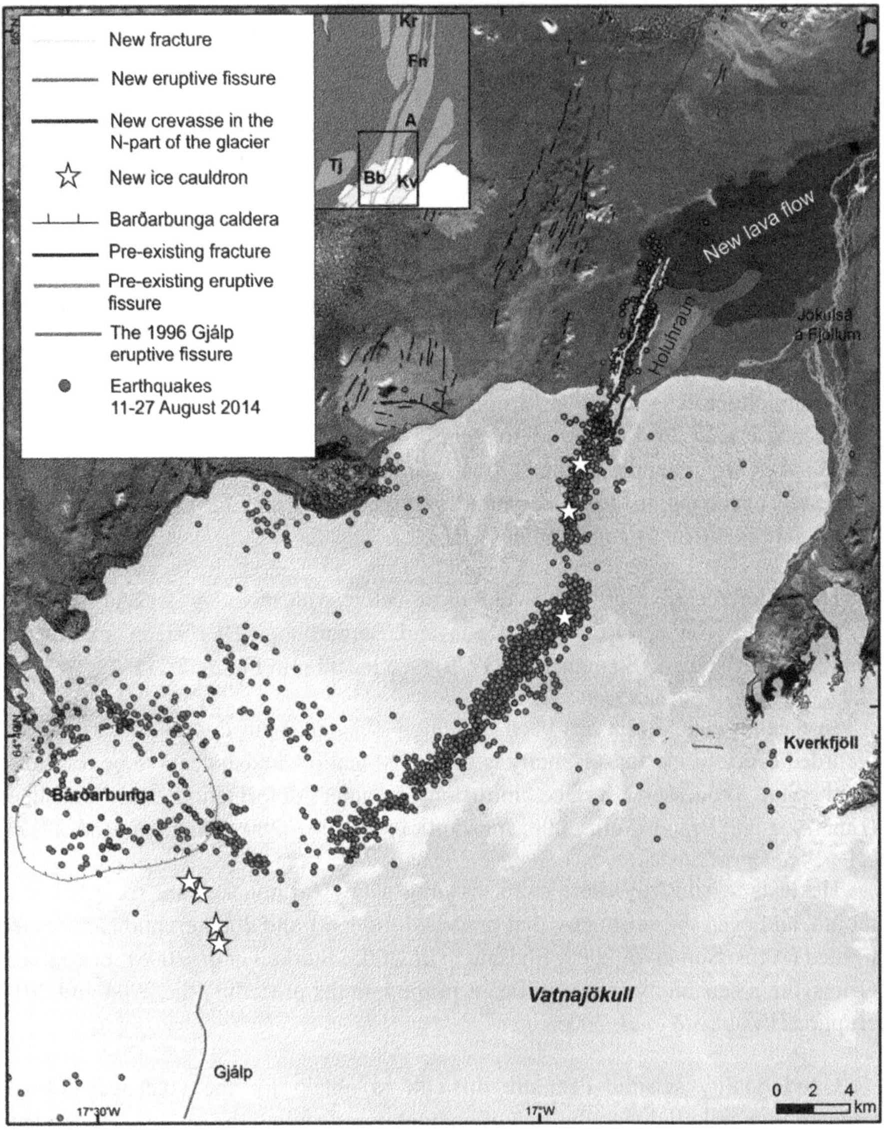

Figure 2.3. *Earthquakes and structures related to the propagation of the Bárðarbunga dyke in 2014 (modified from Hjartardóttir* et al. *2016a). Map: Location of the Bárðarbunga in the southern part of the NVZ . A: Askja; Bb: Bárðarbunga; Fn: Fremri Námar; Kr: Krafla; Kv: Kverkfjöll; Tj: Tungnafellsjökull*

Part of the seismicity of volcanic origin is linked to the propagation of dykes. An extremely well-documented example is of Bárðarbunga in August 2014 as it has been monitored and analyzed in detail (Guðmundsson *et al.* 2014; Sigmundsson *et al.* 2015; Hjartardóttir *et al.* 2016a).

From August 11 to 27, 2014, a 47-km-long dyke propagated laterally from this central volcano, generating 17,000 earthquakes in 2 weeks, i.e. more than the whole records in Iceland in a normal year. The dyke initiated from the central volcano and gradually extended northeast and then north-northeastward, north of latitude 64°43 (Figure 2.3). This change in direction, from N 45° E to N 15° E, corresponds to the general directions of the EVZ and NVZ. It is identical, at the same latitude, to the change in direction of the dykes on the eastern coast of Iceland, suggesting that it has persisted for about 10 million years (Guðmundsson *et al.* 2014).

An effusive eruption followed for 6 months, during which the caldera gradually collapsed with a subsidence of 65 m (Sigmundsson *et al.* 2015), causing more than 80 earthquakes of magnitude M 5. The seismicity decreased at the end of February 2015 when the eruption and collapse of the caldera stopped. It resumed 6 months later while GPS data indicated a new swelling of the volcano. Since then, many earthquakes – some of them up to magnitude M 4.5 – have been recorded. They are consistent with a recharge of the volcano.

2.2.1.2. *Seismicity and tectonics*

Several thousand earthquakes are recorded each year by the SIL network. For example, in the period 1994–2007, more than 250,000 earthquakes occurred in Iceland and the surrounding offshore platform (Figure 1.6), with a magnitude between –2 and 6.6. About 10,000 of them had a magnitude $M_L > 2$. More than 80% of them were located in the TFZ (48,800 earthquakes), the SISZ (35,200), the Hengill sector (99,200) and the Reykjanes Peninsula (RP) (18,900) (Jakobsdóttir 2008).

2.2.1.2.1. Seismicity of the TFZ

While it is generally accepted that the TFZ consists of three seismic lineaments (Figure 2.4), the Dalvík Line (DL), the Húsavík-Flatey Fault (HFF) and the Grímsey Line (GL), only the HFF and GLs are currently the site of significant microseismicity. This has been extensively described by several authors, in particular Rögnvaldsson *et al.* (1998) and Riedel *et al.* (2005). Most of the microseisms in the TFZ occur on N-S to NNE-SSW faults characterized by left-lateral motion (Rögnvaldsson *et al.* 1998; Stefánsson *et al.* 2008) with the exception of those related to the HFF where the fault planes are dextral and ESE-WNW.

Historical earthquakes in the TFZ are not as well documented as they are in the SISZ, and their traces are not easily characterized there. Thus, they will be discussed here together with the current earthquakes and not in section 2.3.2.1, which refers to seismic faults analyzable in the field.

Despite its present-day low microseismic activity, the occurrence of recent major earthquakes has been demonstrated along the DL, in the Tröllaskagi and Flateyjarskagi peninsulas and at sea (Stefánsson *et al.* 2008) (Figure 2.4). For example, in 1934, an earthquake of magnitude M 6.2 (Karnik 1969) occurred on the east coast of the Tröllaskagi peninsula, destroying several buildings in Dalvík city. The recent microseismic activity suggests left-lateral motion on a NNE-SSW oriented fault, consistent with the northward propagation of rupture zones during the 1934 earthquake (Þórarinsson 1937; Långbacka and Guðmundsson 1995; Rögnvaldsson *et al.* 1998; Stefánsson *et al.* 2008).

The earthquake of magnitude M 7 that occurred at the mouth of the Skagafjörður in 1963 can also be connected to the activity of the DL. Strong aftershocks (M_b 4-5) occurred in an elongated area in the direction of the KR. The nodal plane selected as the seismic fault by Stefánsson (1966) was NNE-SSW and sinistral, which is in good agreement with the alignment of current microseisms (Stefánsson *et al.* 2008).

However, the significant micro- and macroseismic activity of the TFZ is mainly concentrated, at the moment, on the other two lines: the HFF and the GL.

Most seismic faults in the western part of the HFF are N 122°–140° E and dextral. In its eastern part, there are both dextral faults parallel to the transform direction and some N-S sinistral faults (Rögnvaldsson *et al.* 1998). The largest known historical earthquake on the HFF dates from 1755 with an estimated magnitude M_S 7.2 (Halldórsson 2005).

In 1872, another major sequence with two major earthquakes (M 6.5) caused significant damage in the village of Húsavík and on the island of Flatey. At least the first of these two earthquakes was related to the right-lateral motion of the HFF, while the second may have occurred on a perpendicular and sinistral fault (Stefánsson *et al.* 2008). Metzger *et al.* (2011, 2013) estimate that there is currently a seismic potential of magnitude M_W 6.8 on the HFF, posing a significant threat to the city of Húsavík.

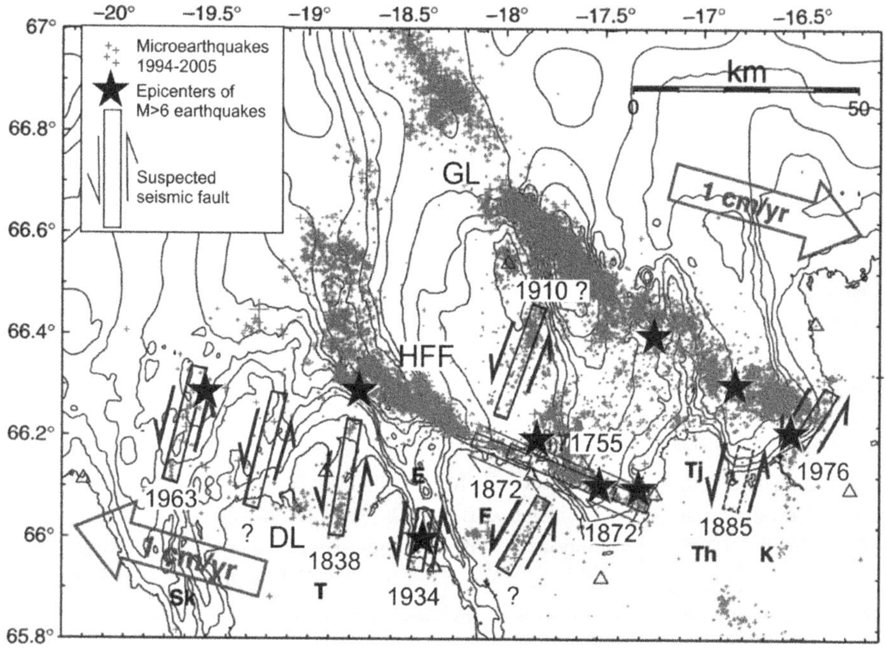

Figure 2.4. *Seismic faults and microseismicity in the Tjörnes Fracture Zone (modified from Stefánsson* et al. *2008)*

COMMENT ON FIGURE 2.4.– *The blue rectangles and associated arrows indicate the probable locations and motions of faults corresponding to historical and recent earthquakes of magnitude 6–7, with their date of occurrence. The locations of epicenters (black stars) are based on the intensities for pre-1963 earthquakes. E: Eyjafjördur; F: Flateyjarskagi peninsula; K: Krafla fissure swarm; Sk: Skagafjörður; Th: Þeistareykir fissure swarm; Tj: Tjörnes peninsula; T: Tröllaskagi peninsula; DL: Dalvík Line; HFF: Húsavík-Flatey Fault; GL: Grímsey Line. The movement of the plates is indicated by the large brown arrows.*

In recent years, most of the microseismic activity on the HFF has occurred in its western part, near its junction with the KR, as well as about 40 km to the east, near its intersection with the Grímsey shoal. The fault planes estimated from earthquake relocations have directions between N 130° E and N 146° E, dips of 70°–90° and their motion is right-lateral. Close to the KR, the seismic faults have a submeridian direction and their motion becomes normal (Rögnvaldsson *et al.* 1998). The faults on the flank of the Grímsey shoal, NNE-SSW direction, are considered normal, but the focal mechanisms indicate an important sinistral component. For Rögnvaldsson *et al.* (1998), the horizontal displacement on these faults is two to three times greater than the vertical one. Finally, further east, the earthquakes characterize an N 123° E trending subvertical fault.

Several earthquakes of magnitude greater than 6 have occurred on the GL since the end of the 19th century. The magnitude of the 1885 earthquake (M 6.3) has been evaluated on the basis of intensities (Halldórsson 2005) and there is no indication of the direction of the seismic fault. Stefánsson *et al.* (2008) assumed it to be NNE-SSW and sinistral by analogy with current microseismicity. The largest earthquake (M_S 7.1) occurred in 1910, but despite an instrumental recording (Halldórsson 2005), the epicenter given by Karnick (1969) is far from the seismic fault proposed by Stefánsson *et al.* (2008) on the basis of intensities.

In recent decades, seismic activity has been particularly significant on the GL, with, in particular, several large earthquake swarms at the end of 1975 and beginning of 1976, simultaneously with the start of the last Krafla rifting period (1975–1984). During this seismic crisis, an earthquake of magnitude M 6.4 occurred on January 13, 1976, with a well-constrained focal mechanism (Einarsson 1979). According to the preferred nodal plane, it is attributed by the authors either to a left-lateral motion on a NE-SW fault (Rögnvaldsson *et al.* 1998), or to a right-lateral motion on a NW-SE fault (Einarsson 1987, 1991). Recent microseismic data tend to favor the sinistral movement (Stefánsson *et al.* 2008).

Recent major earthquakes have occurred mainly at the western end of the TFZ. In February 1994, a magnitude 5.5 earthquake occurred at the GL junction with the KR, followed by a period of several weeks of intense activity in this area. From July 1997 to October 1997, a significant seismic activity also affected the western end of the TFZ with, in particular, an M 4.5 earthquake in July and two M 5 earthquakes in September. Finally, the TFZ experienced significant seismic activity in 2002 (Þorbjarnardóttir and Guðmundsson 2003). A major seismic crisis began on September 16 with an earthquake of magnitude M_b 5.5, whose epicenter was located 53 km northeast of Grímsey Island, in the extreme north of the TFZ (Figure 2.5). It was widely felt along the northern coast of the island. More than 300 aftershocks were recorded, distributed along a 15 km long zone, including one of magnitude M 4.3 on September 17. This was the largest swarm of earthquakes recorded in this zone since 1994 (the year of the extension of the SIL seismic network to northern Iceland). Simultaneously and subsequently, two other swarms have been recorded, 15 km north and 5 km east of Grímsey, including more than 200 earthquakes.

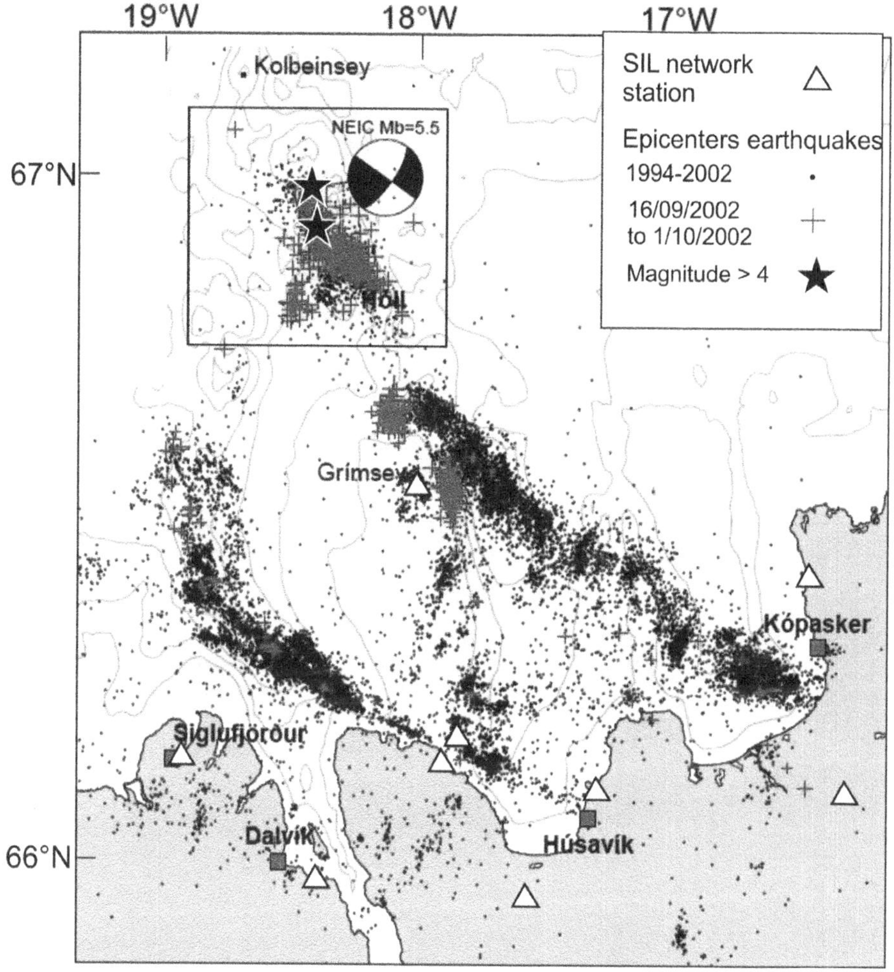

Figure 2.5. *Seismicity in the Tjörnes Fracture Zone (modified from Þorbjarnardóttir and Guðmundsson 2003). Box: Focal mechanism of the earthquake of September 16, 2002 according to the NEIC*

2.2.1.2.2. Seismicity of the SISZ and Reykjanes Peninsula

The SISZ in the south of the island is also affected by significant microseismicity (Figures 1.6 and 2.6). The first Icelandic earthquake whose magnitude (M 7) was instrumentally determined is that of 1912, which was recorded by 18 stations around the world (Karnik 1969). It occurred on the Selsund fault, located at the eastern end of the SISZ (Bjarnason *et al.* 1993; Angelier *et al.* 2004a; Bellou *et al.* 2005). No other major earthquake occurred until the earthquakes of June 17 and 20, 2000 (M_W 6.5), which have been the subject of numerous studies based on surface traces, focal mechanisms and Insar and GPS data (Árnadóttir *et al.* 2001, 2005; Bergerat and Angelier 2001; Pedersen *et al.* 2001, 2003; Angelier and Bergerat 2002; Bergerat and Angelier 2003; Clifton and Einarsson 2005; Antonioli *et al.* 2006; Angelier *et al.* 2008; Daniel *et al.* 2008). This major doublet was followed by several magnitude 5 earthquakes in the RP (Árnadóttir *et al.* 2004).

Based on the knowledge of earthquakes identified since 1700 – and on less accurate information for the earlier period – we may conclude that large earthquakes generally occur during significant sequences affecting a large part of the SISZ. These sequences can last from a few days to several years; each sequence usually begins with a major event in the eastern or central part of the area, followed by smaller events to the west. According to Stefánsson and Halldórsson (1988), the duration of a complete seismic cycle (accumulation and release of stresses) can be as long as 140 years.

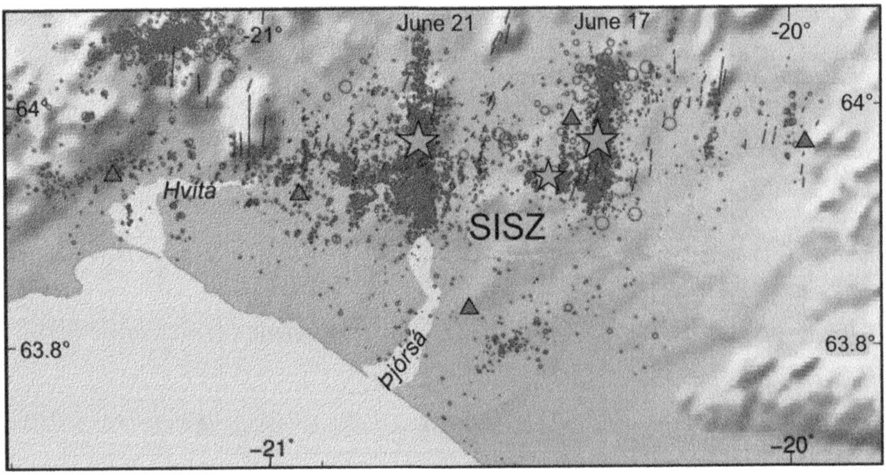

Figure 2.6. *Seismicity in the South Icelandic Seismic Zone (modified from Stefánsson* et al. *2006). Small red circles: Seismic activity between June and December 2000. Green stars: Hypocenters of the major earthquakes of 17 and 21 June 2000 (M_L > 5). Purple triangles: SIL network stations*

In June 2000, about 100 years after the 1896–1912 crisis, the SISZ was affected by a major seismic crisis, including two major earthquakes that were felt within a radius of about 200 km (Figure 2.6). The first, of magnitude M_W 6.5, occurred on June 17 at 15:40 m 41 s UTC and was about 6.3 km deep. This was followed 2 min later, 4 km to the west, by an earthquake M_L 5.7 and in the RP by two earthquakes of M_W 5.5 and M_W 5 (Pagli *et al.* 2003; Árnadóttir *et al.* 2004; Antonioli *et al.* 2006). The second earthquake, also of M_W 6.5, occurred on June 20 at 00:51 m 47 s UTC and was 5.1 km deep. The two major earthquakes, as well as the M_L 5.7 earthquake, occurred on right-lateral strike-slip faults, the Árnes and Kvíarholt faults (June 17) and the Hestfjall fault (June 21) (Figure 2.7).

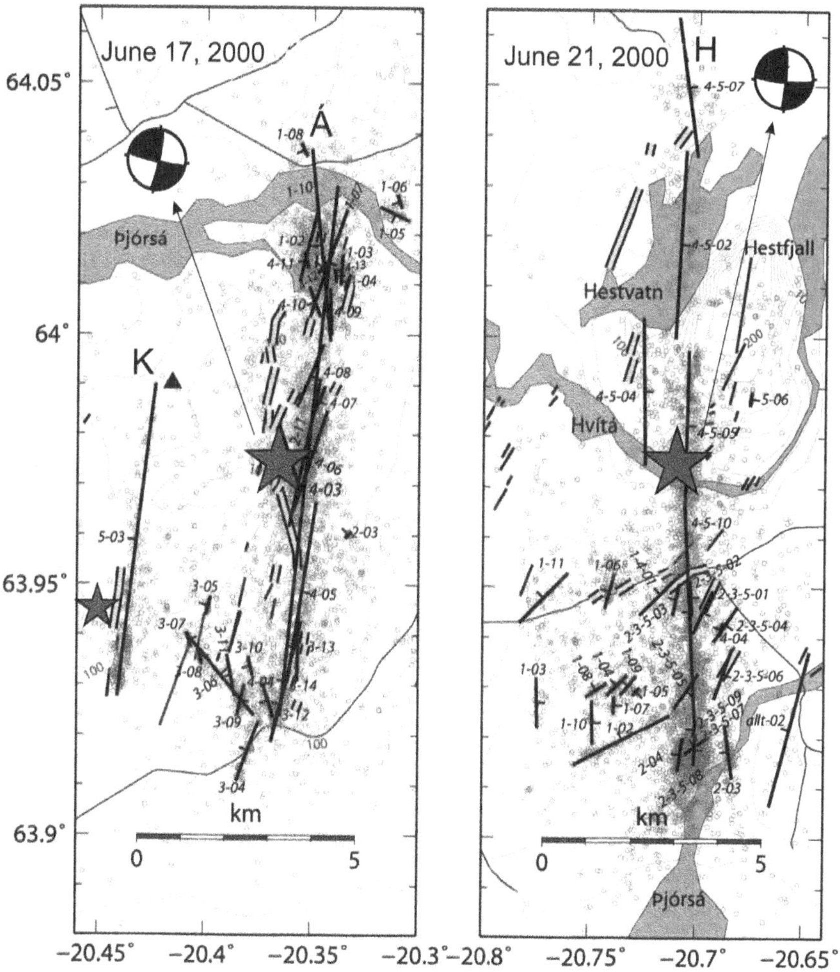

Figure 2.7. *The faults of the earthquakes of June 17 and 21, 2000 (modified from Hjaltadóttir 2009). Black lines: Segments of the Árnes (Á), Kvíarholt (K) and Hestfjall (H) faults. Small gray circles: earthquakes. Red stars: hypocenters of the three main earthquakes. Yellow lines: surface traces (from Clifton and Einarsson 2005). The focal mechanisms of the two M > 6 earthquakes are shown*

Calculations made after the crisis indicated for the June 17 earthquake a 16 km long fault with an azimuth 9° and a dip of 86° E, to 10 km deep, and for the June 21 earthquake a 2° azimuth fault, vertical, 8 km deep and 18 km long (Stefánsson *et al.* 2000). Calculations made later specify these data, giving for the June 17 earthquake a fault with azimuth 7°, 12 km long, and for the June 21 earthquake a fault with azimuth 179°, 16.5 km long (Stefánsson 2006).

Figure 2.8. *Surface traces of the June 17 and 21, 2000 earthquakes*

COMMENT ON FIGURE 2.8.– *(A) Trace of the dextral segment of the June 17, 2000 fault at Mykjunes. (B) E-W sinistral fracture linking two open fractures, near Bitra, on an asphalted car park of Iceland's main road (June 21, 2000). (C)* En échelon *fractures indicating left-lateral movement, east of Bitra (June 21, 2000). (D) Large open fracture, east of Bitra (June 21, 2000). (E) Rising water table, 1 month after the seismic crisis of June 2000, near Bitra (photos A–D: Françoise Bergerat©; photo E: Brigitte Van Vliet-Lanoë©).*

The surface traces associated with this seismic crisis have been mapped at the small scale (Clifton and Einarsson 2005) and show the same characteristic arrangement of step and push-up fractures with regard to the historical faults (section 2.3.2.2).

In some areas where the conservation of traces was remarkable in grassland, the structures at the ground surface have been investigated in detail (Bergerat and Angelier 2001, 2003; Angelier and Bergerat 2002). For example, at the Mykjunes farm, the Árnes Fault shows a system of conjugate N 30° E dextral and N 60° E sinistral strike-slip faults, each formed by arrays of fractures and push-ups (Figure 2.8(A)); the average horizontal offset is 0.25 m for the dextral segment and 0.17 m for the sinistral segment. The dextral submeridian Hestfjall fault is associated with a sinistral fault, the N 50°–60° E trending Bitra fault that includes large *en échelon* fractures, sometimes widely open (Figures 2.8(C) and (D)). The analysis of push-ups (section 2.3.2.2) shows an average horizontal offset of 0.35 m. Two segments of the Hestfjall Fault, studied near Eyvík and Torfagil, show associations of open fractures and normal faults.

All of these observations show that the structures related to these earthquakes are not limited to the main N-S, dextral faults revealed by the focal mechanisms, but also include, on more local scale, extension structures and conjugate strike-slip systems, which are coherent with the general left-lateral movement of the SISZ.

The RP, that represents a connection zone between the RR and the SISZ, is also the seat of significant seismicity. In addition to the three earthquakes that were triggered after the Árnes earthquake of June 17, 2000, a major seismic doublet occurred about 40 km east of Reykjavik on May 29, 2008, with the second shock occurring 3 s after the first (Figure 2.9). The slip occurred on two N-S structures 5 km apart.

Models produced from the joint inversion of GPS and InSAR data indicate that most of the slip of the first event (Ingólfsfjall) occurred at a depth of 2–4 km and that the second (Kross) was located deeper, between 3 and 6 km. Magnitudes were estimated to be M_W 5.8 and M_W 5.9, respectively (Decriem *et al.* 2010). According to Decriem *et al.* (2010), the 2008 doublet was the continuation of the seismic sequence initiated in June 2000. This 2000–2008 sequence would have released about half of the moment accumulated since the 1896–1912 sequence, so it is likely that other such events will occur in the coming decades.

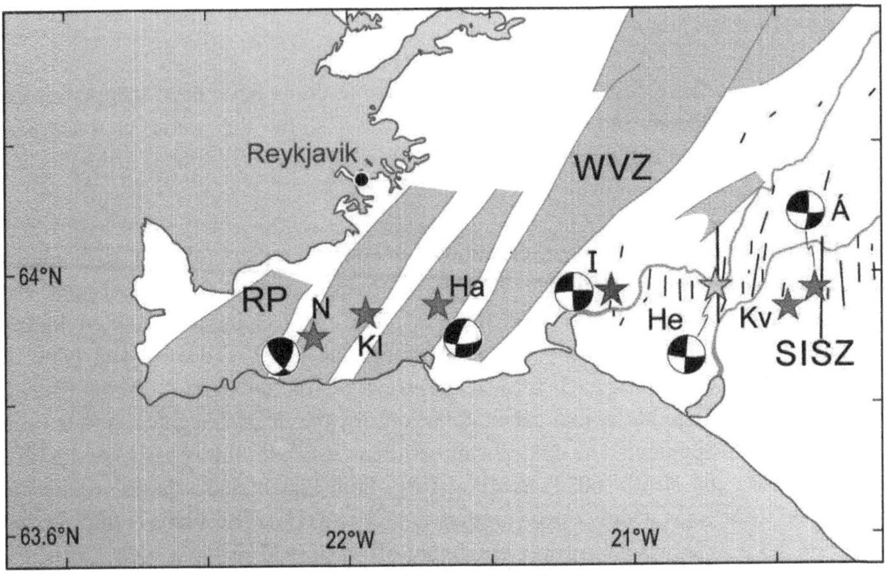

Figure 2.9. *Seismicity in the Reykjanes Peninsula*

COMMENT ON FIGURE 2.9.– *Stars represent the hypocenters of major earthquakes (M_W > 5) (modified from (Árnadóttir* et al. *2004; Decriem* et al. *2010). In green: Earthquakes of 17 June 2000 (Árnes (Á), Hvalhnúkur (Ha), Kleifarvatn (Kl), Kvíarholt (Kv), Núpshlídarháls (N)); in yellow: earthquake of June 21, 2000 (Hestfjall (He)); in red: initial earthquake of May 29, 2008 (Ingólfsfjall (I)). The focal mechanisms of the main earthquakes are shown.*

2.2.1.3. *Interaction between glaciation, tectonics and seismicity*

The Quaternary sedimentary deposits display evidence for imprints of two main types of paleo-seismicity: a high-intensity seismicity that occurred in the interglacial ("temperate") period, as described above, and a recurrent but low-intensity seismicity, associated with more or less significant deglaciation episodes. Ice loading can also lead to the widening of the active rift zone (Bourgeois *et al.* 1998; Bourgeois 2000).

Interglacial seismicity generally occurs in a context where water tables are low (due to basalt fracturing), so shear strength is high and magnitudes are significant.

"Seismic pumping" can, however, cause deep aquifers to resurface for a few weeks, as was the case in June 2000 (Figure 2.8(E)).

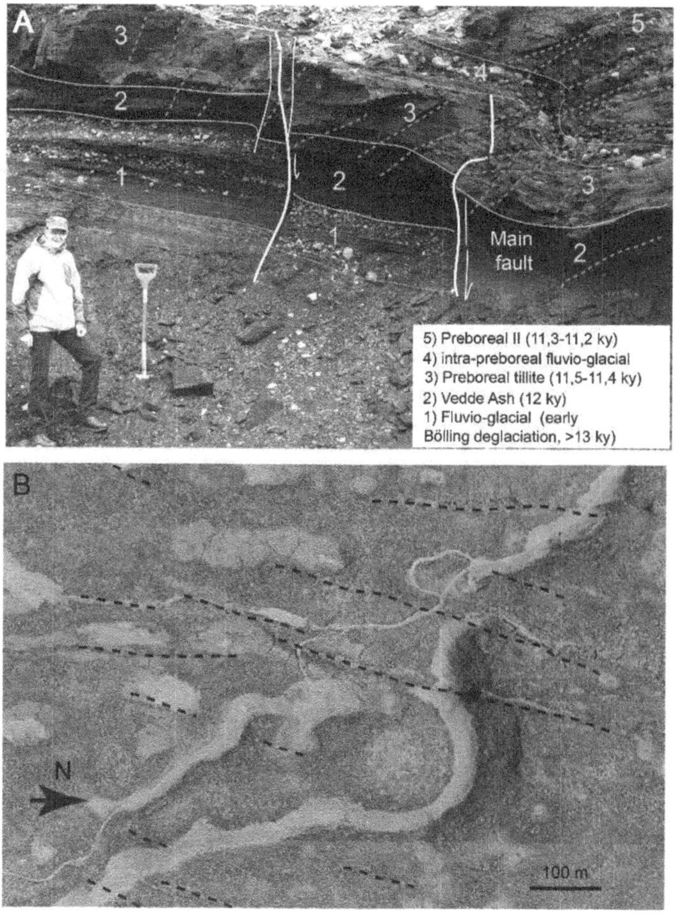

Figure 2.10. *Sedimentary record of fault activity during deglaciation*

COMMENT ON FIGURE 2.10.– *(A) Trace of the Minnivellir seismic fault (section 2.3.2.2) showing a small vertical component (flower structure), accommodating the deglaciation of a Preboreal glacial advance (11.5–11.4 ky), then modified by glacitectonics during a second glacial advance (slip as red dashed lines) at Akbraut (Þjorsá valley) (Brigitte Van Vliet-Lanoë©). Note the collapsed structure close to the fault in the tillite. The glacier progressed to the left. (B) Remobilization of rift faults at the beginning of deglaciation (around 11.2 ky), along the Jökulsá á Fjöllum, cutting a former esker (sub-glacial outfall) from the Krafla massif (aerial view, Maxar Technologies©).*

In the case of the June 17 earthquake, the rising of the water table and thus the presence of fluids may have reduced the number of aftershocks, some of the slips then being aseismic (Pagli *et al.* 2003; Árnadóttir *et al.* 2004). The seismic crisis was followed by partial drainage of Lake Kleifarvatn, in the RP, promoted by fractures in the lake bottom (Clifton *et al.* 2003). The water level dropped by about 4 m during the 18 months following the earthquakes. In general, ice loading tends to extinguish seismic activity (Hasegawa and Basham 1989).

Deflection of the crust due to glacial overload also creates “stress barriers” trapping magma. Magmatism during glacial periods results in the development of magma chambers, whereas eruptions are more common during interglacial periods (Guðmundsson 1986, 2000b). Tremor-type seismicity is expressed during deglaciation with high frequency and low intensity ($<M_L3$) (Van Vliet-Lanoë *et al.* 2005). It is partly induced by the injection of dykes in the rift, in relation to an increase in partial melting induced by the discharge and feeding of subcrustal reservoirs (Guðmundsson 2000b, 2006; Hartley and Thordasson 2013; see Chapter 1 of Volume 2).

In such conditions, water escapes (sand volcanoes), fluid injections (sedimentary dykes) and load casts are the most frequent synsedimentary deformations. Their record occurs usually in waterlogged environments (meltwater). Deglaciation is also accompanied by progressive reactivation of rift faults (Figure 2.10(B)) that intersect glacial sedimentary figures, commonly in association with recurrent microseismicity. During the Quaternary, the reactivation of these faults may have promoted the discrete and polyphased formation of grabens or hemi-grabens. Field observations evidence this along the Barðarbungá-Veiðivötn fracture system (Plateaux 2012) or the Hekla fracture system (Van Vliet-Lanoë *et al.* 2020a).

In the transform zones, the stress level is relatively high and seismicity is strong, especially in the interglacial period. Shear stresses accumulate to be released during earthquakes of magnitude M 6–7; those of 2000–2008 in the SISZ and RP are a contemporary example.

Seismic activity in rift zones is very different from that in transform zones, and parts of the divergent zones can remain seismically silent for long periods of time despite the permanent expansion. Thus, while GPS measurements indicate that the

EVZ absorbed most of the movement during the 2012–2016 period, no significant seismic activity occurred, and the area has experienced only low seismic activity in recent decades (Einarsson 1991). This suggests that although stresses are steadily accumulating, their level is not yet sufficient to cause significant seismicity (Sigmundsson *et al.* 2018b). It should also be noted that part of the tectonic activity is undoubtedly aseismic.

2.2.2. Motions at plate boundaries

Direct confirmation of the spreading rate at the plate boundary in Iceland has been provided since 1993 by the ÍSNET GPS geodetic campaigns (Árnadóttir *et al.* 2009; Sigmundsson *et al.* 2018a). The horizontal velocity field shows, on a small scale, not only the plate spreading, but also, *pro parte,* the glacial isostatic adjustment (GIA) related to the retreat of the Icelandic ice cap since 1890 and responsible for an uplift of the central part of Iceland greater than 30 mm/year. The models developed (Árnadóttir *et al.* 2009), considering the thinning of the ice during this period, show indeed a horizontal component of the GIA that can reach a few millimeters per year, consistent with the observations.

The motion of Iceland relative to the supposedly stable Eurasian plate, determined from the ÍSNET data (Figure 2.11), is roughly identical to that predicted by the MORVEL model (DeMets *et al.* 2010): 18–19 mm/yr in a direction N 100°–105° E, accommodated to the north by a single rift zone (NVZ) and to the south by the overlapping rift zone (WVZ-EVZ) and the RP. Numerous continuous GPS measurements, providing a three-dimensional time series of displacements for more than 20 years (Geirsson *et al.* 2006, 2010), confirmed these measurements.

In detail, because the southern part of Iceland is structurally more complex than the northern part, several studies have attempted to assess the respective spreading rates of the WVZ and EVZ. For example, Sigmundsson *et al.* (1995) estimated that, at present, only 15 ± 15% of the expansion is accommodated by the WVZ, the EVZ, which moreover would propagate southwards, thus absorbing 85 ± 15% of it. Based on more recent data, Decriem and Árnadóttir (2011) found a velocity of 19 mm/year in the SISZ resulting in a negligible rate of expansion at the WVZ. Travis (2012) included the Hreppar block as an independent microplate in his model and concluded that 12–23% of the motion was accommodated in the WVZ. In any case, despite the differences between the proposed models, it is clear that the SISZ and the EVZ currently play a major role as plate boundaries in the south of the island.

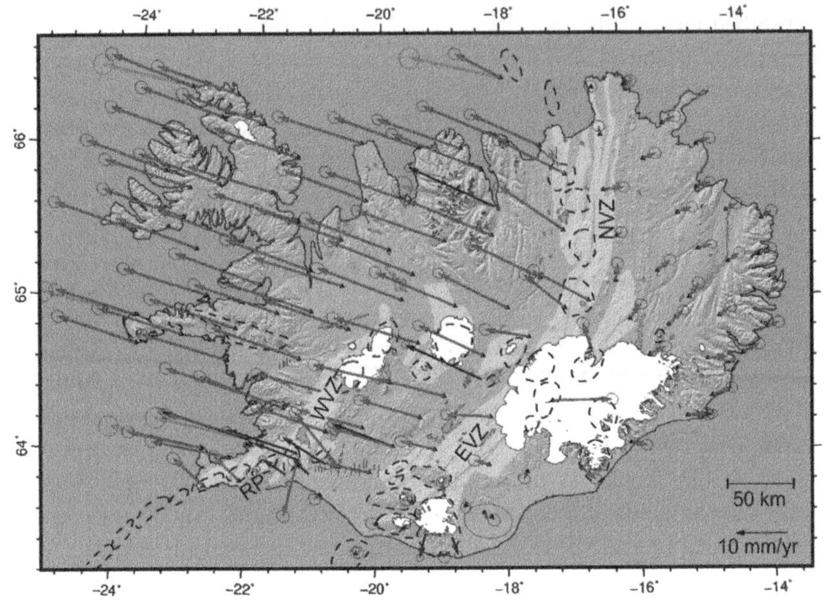

Figure 2.11. *Plate boundary motions in Iceland (modified from Sigmundsson* et al. *2018a)*

COMMENT ON FIGURE 2.11.– *Horizontal velocities of GPS stations relative to stable Eurasia, determined from ÍSNET measurements for the period 1993–2004 (red arrows) and CGPS stations in Iceland for the period 1999–2004 (blue arrows): 95% confidence ellipses. The green arrows indicate the relative motion of the North American plate with respect to stable Eurasia, based on the MORVEL model (DeMets* et al. *2010). RP: Reykjanes Peninsula; EVZ: East Volcanic Zone; NVZ: North Volcanic Zone; WVZ: West Volcanic Zone.*

However, it should be kept in mind that GPS displacement measurements only concern a very limited time scale, a "snapshot" of the deformation, which does not necessarily reflect what is happening on a longer time scale in the WVZ and EVZ (section 2.3.1.2).

2.3. Iceland's main structural features

Iceland's present-day structural model with the three active rift segments and the two transform zones (Figure 2.2) is partly the result of its heritage and mainly of the tectonic reorganization following the rift jumps that occurred during the last 10 My.

2.3.1. *The paleo-rifts and the active rift*

2.3.1.1. *The paleo-rifts*

We have seen (section 2.1) that during the westward migration of the Mid-Atlantic Ridge, the Icelandic rift became progressively inactive and a new rift was created above the apex of the mantle plume (Figure 2.1(B2)).

This rift jump phenomenon implies the abandonment of the paleo-rift and the initiation of a new one. Even if the rift jump model is accepted, the process itself is poorly constrained, as are the exact location of the paleo-rifts, their age or the duration of the concomitant functioning of the two rifts, the old and the new. The initiation of a new rift as well as the temporary resumption of a paleo-rift could be linked to pulses of the mantle plume (Walters *et al.* 2013). Lava flows may extend over several tens of kilometers, and the last flows may thus seal the older tectonic structures. Consequently, it is difficult to locate precisely the paleo-rifts, except by guessing their location based on morphology. Indeed, in active rift zones, lava piles are inclined toward the rift axis, thus forming a synform in the landscape. Finding the ancient rift zones thus consists, first of all, of searching for the synforms in the ancient lava piles.

In the early 1970s, Ward (1971) and Sæmundsson (1974) located the paleo-rifts according to the age of the lavas (mostly K/Ar analyses), the paleomagnetic data and the structural observations ("ages discontinuities" marked in the field by angular discordances of the lava layers and underlined by an erosive surface).

In the northern part of the island, the last paleo-rift was thus classically located in the Húnaflói-Skagi peninsula (Sæmundsson 1974). The discontinuities inside the lava pile on either side of the present rift have been used to date the rift jump, but without enabling the determination if the extinction of the paleo-rift was synchronous with the initiation of the new rift or if the two rifts were active together and for how long.

Based on the ages available in the literature (McDougall and Wensink 1966; Aronson and Sæmundsson 1975; McDougall *et al.* 1976; Musset *et al.* 1980; Sæmundsson *et al.* 1980; Jancin *et al.* 1985), the Húnaflói-Skagi Rift would have kept active until 1–2 My. The initiation of the new rift, through the 9–10 My old lavas of the eastern flank of the palaeo-rift, would date from 6–7 My, the two rifts having thus functioned simultaneously for 4–6 million years.

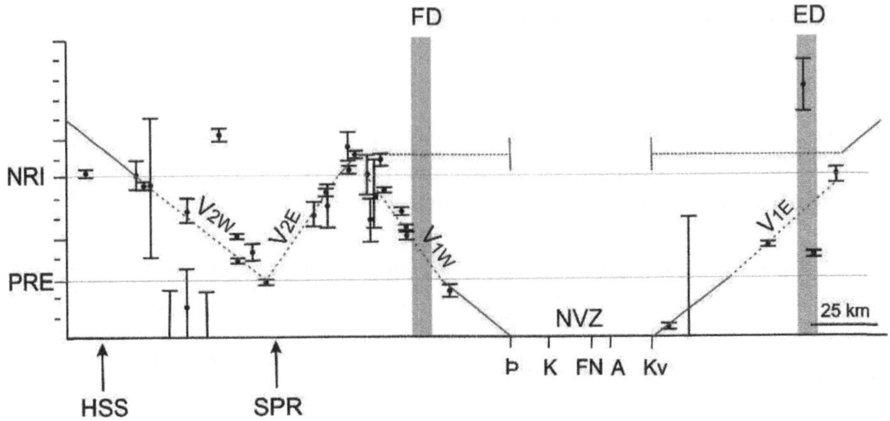

Figure 2.12. *Ages of the (projected) dykes along a transect N 106° E ($^{40}Ar/^{39}Ar$ dating) through the Húnaflói-Skagi synform, the active rift and the Flateyjarskagi and East unconformities (modified from Garcia* et al. *2003)*

COMMENT ON FIGURE 2.12.– *Error bars on ages are shown. Continuous line: period during which only one rift is active; dashed line: period with two active rifts. Plate divergence half-velocities in km/My: V_{2W} paleo-rift (9.2 ± 1.1); V_{2E} (5.3 ± 0.9); V_{1W} new rift (5.7 ± 2.1) and V_{1E} (7.5). PRE: paleo-rift extinction time; NRI: new rift initiation time; SPR: Skagafjörður paleo-rift; HSS: Húnaflói-Skagi synform; ED: East unconformity; FD: Flateykjarskagi unconformity. Central volcanoes of the Northern Volcanic Zone (NVZ) projected: Askja (A), Fremri Námar (FN), Krafla (K), Kverkfjöll (Kv) and Þeistareykir (Þ).*

A study based on $^{40}Ar/^{39}Ar$ dating of the dykes (Volume 2, Chapter 2), was carried out in the early 2000s along an ESE-WNW profile (i.e. parallel to the direction of plate divergence), 350 km long, and intersecting the supposed Húnaflói-Skagi paleo-rift, the present rift and unconformities (East unconformity and Flateyjarskagi unconformity; Figure 2.12). Based on this study, the paleo-rift axis would not be located in the Húnaflói-Skagi Peninsula, but rather 60 km further east, in the Skagafjörður (Garcia *et al.* 2003). This Skagafjörður paleo-rift is thought to have remained active until about 3 My. The onset of the NVZ activity started toward 8–8.5 My; these two rifts would therefore have been functioning simultaneously from 5 to 5.5 My. During this period, the accretion rates of the two rifts were almost equal, but with an asymmetric opening and higher rates on the outer flanks (Garcia *et al.* 2003). It should be noted, however, that no synforms are visible in the Skagafjörður paleo-rift.

Postulating that the existence of a synform in the Húnaflói-Skagi axis is indeed the mark of a paleo-rift leads to considering that this one functioned before the Skagafjörður paleo-rift became active. However, no radiometric dating confirms this hypothesis and there is no discontinuity between the supposed paleo-rift and the Skagaheiði Peninsula in the east (Garcia *et al.* 2003) or the Vestfirðir Peninsula in the west (Moorbath *et al.* 1968; Bagdasaryan *et al.* 1976; McDougall *et al.* 1984).

Garcia *et al.* (2008) proposed, from structural and morphological data, a structural evolutionary pattern following the last rift jump in northern Iceland. This model implies that Tertiary lava flows erupted along the paleo-rift of Skagafjörður have been down-bent under the weight – and in the direction – of the Plio-Pleistocene lava flows emitted from the NVZ and the central part of Iceland. This down-bending process, enhanced by the ice cap load, led to the development of local flexures. Such a structural reorganization could explain the existence of the Húnaflói-Skagi synform, in agreement with radiometric dating, without the necessity to suppose along it the occurrence of a paleorift axis.

An even older paleo-rift is supposed to have existed at sea, off the west coast of the northwestern peninsula (Vestfirðir) where lavas of the Skagafjörður paleo-rift unconformably overlie northwestward-dipping lavas (McDougall *et al.* 1984). However, the ages available to date the rift jump, on either side of this unconformity, do not provide a very coherent pattern (Moorbath *et al.* 1968; Bagdasaryan *et al.* 1976; McDougall *et al.* 1984). $^{40}Ar/^{39}Ar$ datings (Harðason *et al.* 1997) indicate an age of 14.85 ± 0.12 My below the discordance and 15.05 ± 0.08 My above, which would correspond to a very short hiatus.

In the south of Iceland, the last rift jump is easier to date – and locate – since it is still in progress. The WVZ has been active since about 7 My (Smith 1967; Moorbath *et al.* 1968; Bagdasaryan *et al.* 1976; McDougall *et al.* 1977) and the initiation of the EVZ dates back to 2–3 My (Aronson and Sæmundsson 1975; Jóhannesson *et al.* 1990). No unconformity marks the contact of the lavas from each of these zones, and only the Hreppar antiform (Figure 2.13) shows the divergence in the dip of the lavas toward the EVZ and toward the WVZ.

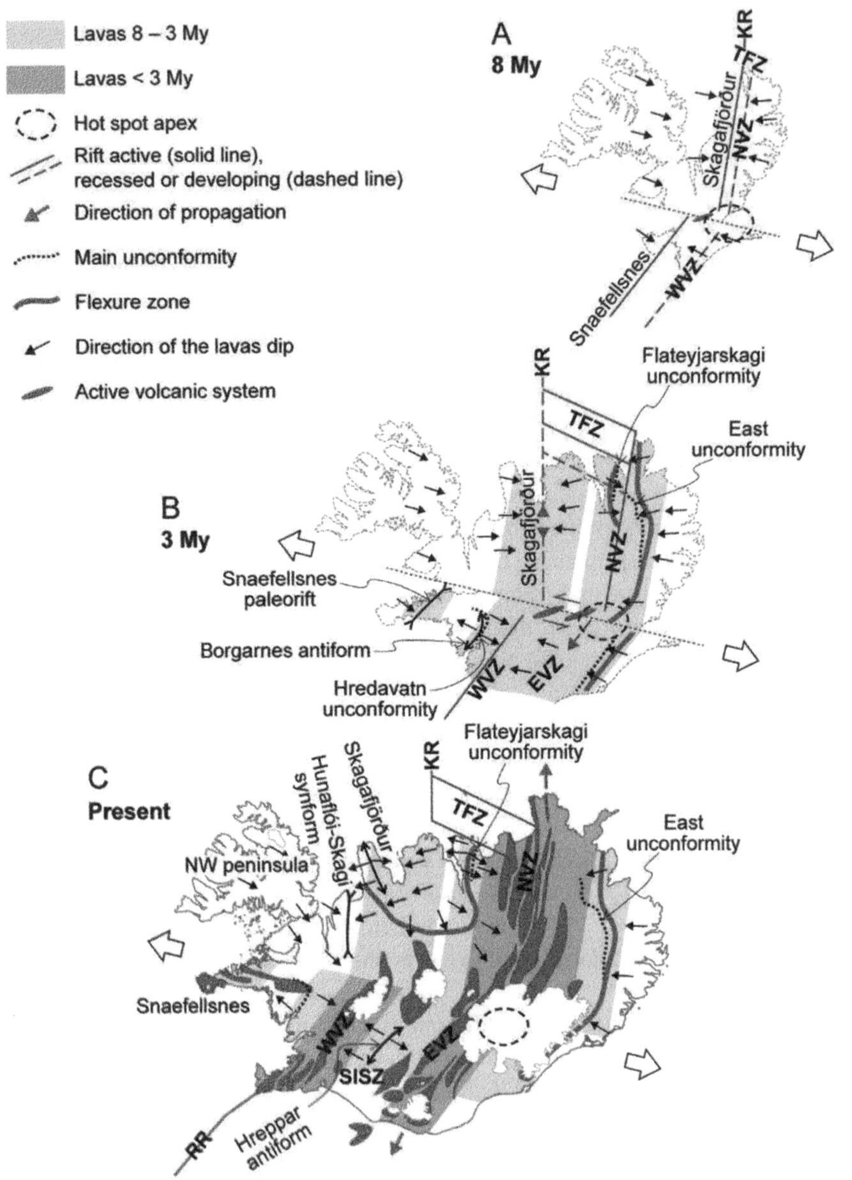

Figure 2.13. *Tectonic evolution of Iceland since 8 My (according to Garcia 2003). KR: Kolbeinsey Ridge; RR: Reykjanes Ridge; TFZ: Tjörnes Fracture Zone; SISZ: South Iceland Seismic Zone; EVZ: East Volcanic Zone; NVZ: North Volcanic Zone; WVZ: West Volcanic Zone. The coastlines in A and B are shown only to facilitate the reading of the figure*

To the west of the WVZ , the lavas of this zone overlie lavas aged 9.4 ± 0.7 My (Moorbath *et al.* 1968; Aronson and Sæmundsson 1975). The latter were emitted within another paleo-rift (Hreðavatn unconformity; Figures 2.13(A) and (B)), whose axis could correspond to the NE-SW oriented Snæfells synform (*Snæfellsnes*). This very ancient rift (about 15 My) would have been active until 6–7 My (Moorbath *et al.* 1968; Harðason *et al.* 1997), the time of the initiation of the WVZ.

2.3.1.2. *The active volcanic zones*

The segments of the active rift in Iceland are called, in all the international literature, NVZ, EVZ and WVZ, respectively. We will thus use these names here, but it should be kept in mind that they concern rifts, and therefore volcano-tectonic zones.

The active volcanic zones (Figure 2.14) include two sets that are distinguished by their petrographic composition: (i) the active rift, including NVZ , WVZ , EVZ, Hofsjökull swarm and RP, with lavas of tholeitic to picritic types, and (ii) the volcanic zones located outside the rift axis, known as "off-rift flank zones" (Sæmundsson 1979; Jóhannesson and Sæmundsson 1998), characterized by alkaline or transitional lavas (Jakobsson 1979). This group is composed of the Snæfells Peninsula (*Snæfellsnes*), the Esjufjöll-Öræfajökull volcanic systems located to the east of the EVZ, and its southern extension (see Chapter 1 of Volume 2).

2.3.1.2.1. Meaning of "flank zones"

The so-called "flank zones" are located outside the volcano-tectonic zones (Figure 2.14), but include Iceland's most active volcanoes, in particular, in the southern extension of the EVZ, Hekla, Torfajökull, Tindfjöll, Eyjafjallajökull, Katla and the Westman Islands. This area is considered to represent a southward propagation of the rift.

The eastern flank zone of the EVZ, composed of the Snæfell, Esjufjöll and Öræfajökull volcanoes, is little known because it is mostly subglacial.

Finally, the Snæfellsnes volcanic zone includes the three volcanic systems of Snæfellsjökull, Lýsuskarð and Ljósufjöll whose lavas cover the Tertiary lavas of the paleo-rift. Several interpretations have been proposed to explain the existence of this zone: a rift branch reactivated since 2 My propagating eastward (Guðmundsson 2000b), a zone of residual activity of the hot spot (Steinthorsson *et al.* 1985) or a transfer zone continuing toward Langjökull and Hofsjökull north of the WVZ (Ward 1971; Sæmundsson 1978).

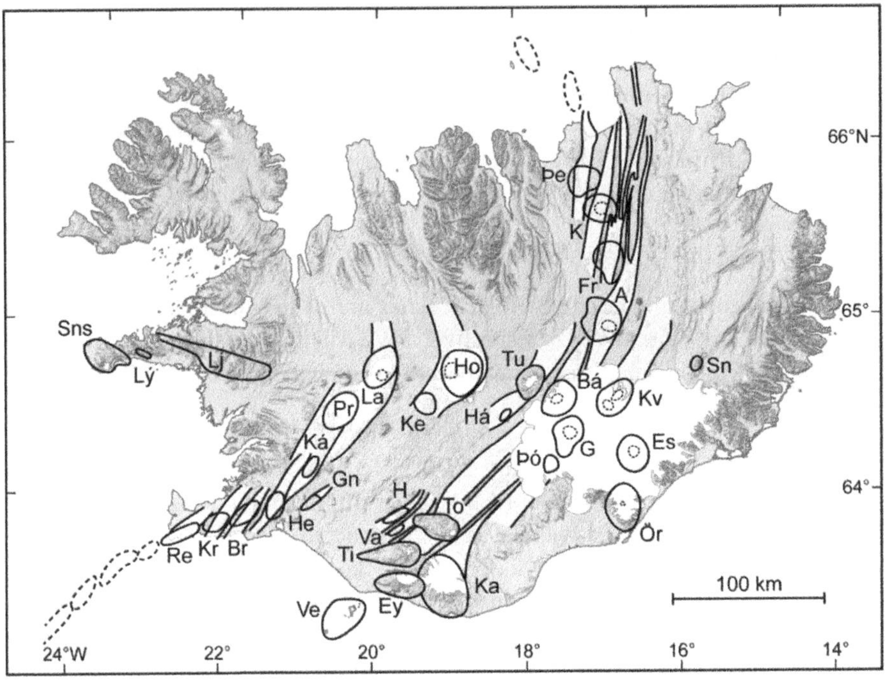

Figure 2.14. *Volcanic systems of Iceland (modified from Einarsson and Sæmundsson 1987; Sigmundsson* et al. *2018b)*

COMMENT ON FIGURE 2.14.– *The volcano-tectonic zones constituting the active rift (NVZ , WVZ and EVZ) are constituted of* en échelon *fissure swarms associated with central volcanoes: Askja (A), Bárðarbunga (Bá), Brennisteinsfjöll (Br), Fremri Námar (Fr), Grímsnes (Gn), Grímsvötn (Gv), Hágöngur (Há), Hengill (H), Hofsjökull (Ho), Kálfstindar (Ká), Kerlingarfjöll (Ke), Krafla (K), Krýsuvík (Ks), Kverkfjöll (Kv), Langjökull (La), Prestahnúkur (Pr), Reykjanes (R), Snæfellsjökull (S), Tungnafellsjökull (Tu), Vatnafjöll (Va), Þeistareykir (Þe) and Þórðarhyrna (Þó). The other volcanic areas ("flank zones") are (i) the Snæfells Peninsula with Ljósufjöll (Lj), Lýsuskarð (Lý) and Snæfellsjökull (Sns), (ii) the southern extension of the EVZ with Eyjafjallajökull (Ey), Hekla (He), Katla (Ka), Tindfjöll (Ti), Torfajökull (To) and the Westman Islands (Ve) and (iii) the Snæfell (Sn)-Esjufjöll (Es)-Öræfajökull (Ör) system east of the EVZ.*

2.3.1.2.2. Special features of the active rift

The Icelandic rift presents notable structural differences compared to the submerged parts of the ridge:

– the absence of a well-marked axial valley. While the submarine morphology of the North-MAR is characterized by an axial valley 1–20 km wide bordered by two "shoulders" that dominate it by about 1.5–3 km, there is no major difference in relief in Iceland between active and inactive volcanic areas. The island presents a globally flattened dome-shaped morphology. In fact, rift shoulders do exist but, due to glacial erosion, they are not topographically marked;

– the dips of the lava piles. These generally converge toward the active volcanic zones (Sæmundsson 1979); this arrangement is different from those of the outwardly tilted blocks that are usually observed on the shoulders of slow ridges;

– the width of the "volcanic zones" – on average about 50 km wide – which distinguishes them from the classic active slow ridge volcanic zones whose width does not exceed 5 km. If considered in its entirety, the Holocene Icelandic volcanic zone (Figures 2.2 and 2.14) has a width of nearly 200 km.

An accretion model was proposed by Bourgeois *et al.* (2005) to explain the width of the deformation zone, the absence of an axial valley, the dip of the lavas and the outcropping extinct magmatic chambers. It implies the activation, growth and decline of a rollover fault system over shallow magmatic chambers (see Chapter 3, Figure 3.21).

Because it formed recently (2–3 My ago), the active eastern rift is nowadays the source of most volcanic production and it accommodates most of the extension in southern Iceland. Several authors believe that the EVZ will eventually replace the WVZ as the single plate boundary in southern Iceland (Pálmason 1981; Meyer *et al.* 1985; Einarsson 1991). In this scenario, the WVZ represents a ceasing volcanic activity area. It is also commonly accepted that the volcanoes of Eyjafjallajökull, Katla, Tindfjöll and the Westman Islands correspond to a southward development of the EVZ, possibly leading to a future southward migration of the transform zone.

However, when considering the overall post-glacial volcanic production, while there is evidence of southward propagation of the EVZ, there is no evidence of a gradual decline of the WVZ. There is no indication of a steady variation in volcanic productivity along the axis of the WVZ during the postglacial period, as would be expected under the assumption of an extincting rift (Sinton *et al.* 2005). Likewise, the low volcanic production in the Þingvellir region is more than compensated by high production to the northeast where the Hofsjökull, in particular, represents a very active volcanic center. This feature is difficult to reconcile with a weakening of the WVZ (Sinton *et al.* 2005).

2.3.1.2.3. Structures of the active rift

The different branches of the rift, although they differ in detail, have in common that they are composed of so-called fissure swarms arranged more or less *en échelon* (Sæmundsson 1978), 5–20 km wide and 40–100 km long, where open fractures – eruptive or not – and normal faults coexist (Figure 2.15). Each fissure swarm is associated with a central volcano (Sæmundsson 1974) (Figure 2.14).

Fissure swarms are the direct volcano-tectonic manifestation of the rifting process and thus allow its structure to be specified as well as the dynamics of its initiation and development. Volcanic activity leads to the emplacement of numerous dykes and the sealing of fractures by lava flows at variable time intervals. The surface expression of the fractures has been studied in great detail in a number of remarkable sectors of the Icelandic rift, such as the Þingvellir fissure swarm in the WVZ (Sæmundsson 1974; Guðmundsson 1987b; Acocella *et al.* 2000; Grant and Kattenhorn 2004; Sonnette *et al.*, 2010) and the Vogar fissure swarm (VFS) in the RP (Guðmundsson 1980, 1986, 1987a; Forslund and Guðmundsson 1991; Grant and Kattenhorn 2004; Villemin and Bergerat 2013).

Figure 2.15. *Krafla fissure swarm in the Leirhnjúkur sector, view toward the south-southeast (Françoise Bergerat©)*

COMMENT ON FIGURE 2.15.– *The aa (black) lavas from the 1984 eruption are visible on the right of the photograph and the pahoehoe (gray) lavas from the 1729 eruption on the right in the background. The surface covered with vegetation and cut by faults and open fractures is that of older lava, dating from the beginning of the Holocene, resting over hyaloclastites. The Krafla caldera is thought to date from the last interglacial (K-Ar dating) (age of lavas according to Sæmundsson* et al. *2012).*

Figures 2.16 and 2.17 illustrate the example of the VFS and its analysis based on a photogrammetric model, using GPS ground benchmarks for calibration and complementary field observations (Villemin and Bergerat 2013).

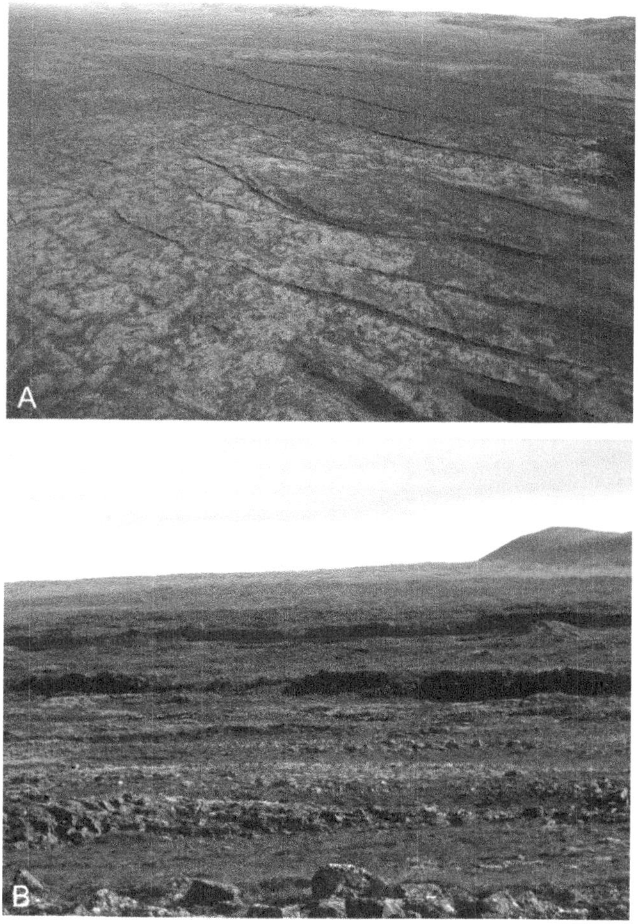

Figure 2.16. *West facing normal faults illustrating the step pattern in the Vogar fissure swarm, northern part of the Reykjanes fissure swarm (post-glacial lavas). (A) Oblique aerial photograph taken from a helicopter (view toward the east, Thierry Villemin©). (B) Detail of two steps (view toward the southeast, Françoise Bergerat©)*

Each fracture is thus identified and positioned in space in three dimensions (Figure 2.17(A)), which also allows automatic transverse (Figure 2.17(B)) and longitudinal (Figure 2.24) topographic profiles to be drawn.

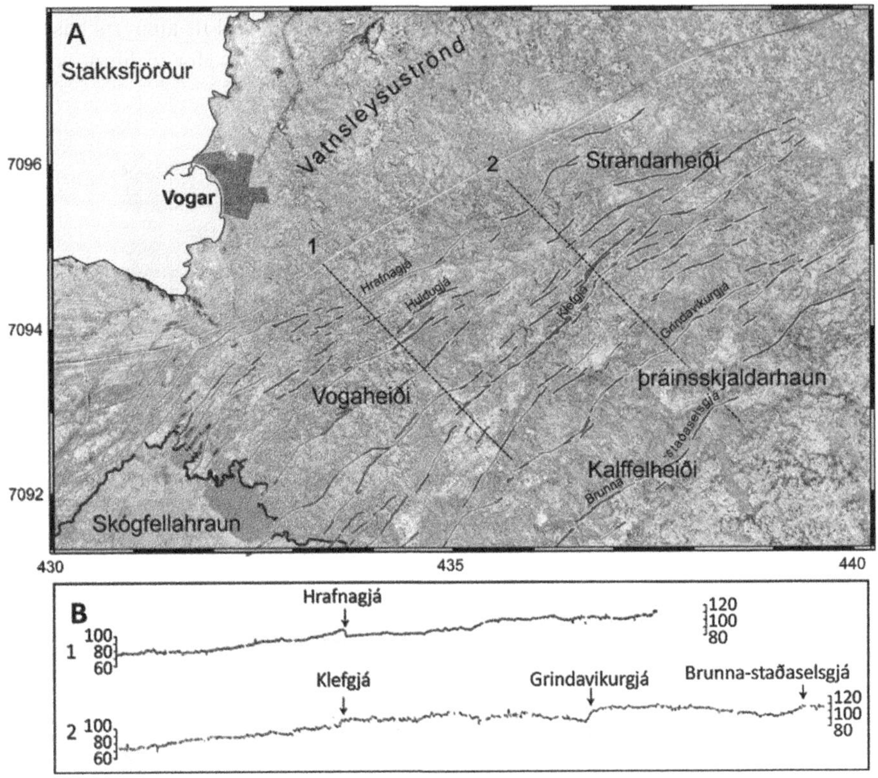

Figure 2.17. *The Vogar fissure swarm, Reykjanes Peninsula (modified from Villemin and Bergerat 2013)*

COMMENT ON FIGURE 2.17.– *(A) Map created from photogrammetric models created from a mosaic of aerial photographs. Normal faults dipping eastward, in blue (in purple: open); normal faults dipping westward, in green (in yellow: open); fissures, in red. Location grid in UTM coordinates. (B) Examples of two transverse profiles across the fissure swarm . Lengths and altitudes are given in meters.*

The general structure of the VFS is, in the south, that of a graben and in the north that of a hemi-graben. The fracturing includes both fissures (without vertical offset) and normal faults, all generally accompanied by an opening (ranging from a few millimeters to several meters). Most fractures form composite lines, generally including both fissures and normal faults (Figure 2.17(A)). Maximum offset on normal faults is about 10 m and appears to be independent of the total length of the faults.

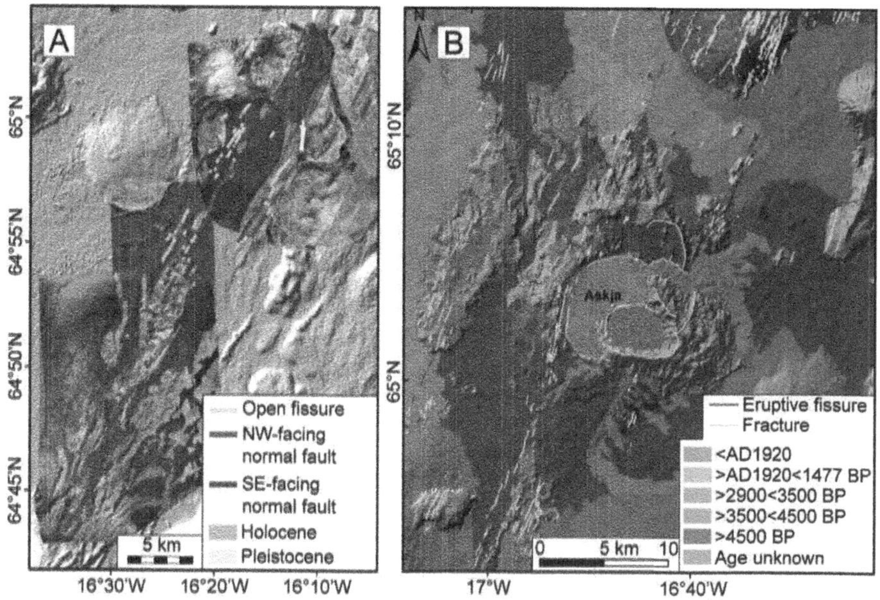

Figure 2.18. *Fissure swarms in the NVZ . (A) Kverkfjöll (modified from Plateaux 2012). (B) Askja (modified from Hjartardóttir* et al. *2016b). BP (before present) means "before 1950" and AD (anno domini) "year 1 of the Christian calendar"*

Several other fissure swarms have also been analyzed in detail in the NVZ, such as the Krafla fissure swarm (Opheim and Guðmundsson 1989; Guðmundsson and Bäckström 1991; Angelier *et al.* 1997; Dauteuil *et al.* 2001) and those of Askja and Kverkfjöll (Figure 2.18), also mapped by photogrammetric means associated with field verification campaigns (e.g. Clifton and Kattenhorn 2006; Hjartardóttir *et al.* 2009; Plateaux 2012; Villemin and Bergerat 2013; Einarsson *et al.* 2018). Detailed studies are less common in the EVZ (Plateaux 2012). In the majority of cases, fissure swarms are organized in more or less marked, often asymmetrical, graben systems.

The main feature of fracturing in the Icelandic rift is the deformation partitioning between normal faults, open fractures and fractures injected by volcanic material.

Normal faults, located at the edge or within the fissure swarms, have two characteristics on outcrops: (i) they are subvertical and (ii) they are open (Figure 2.19(A)). A small graben, bounded by the main fault, and an antithetical secondary fault may also sometimes be present (Figure 2.19(B)). It is easy, from this surface geometry , to reconstruct the geometry of these faults at depth.

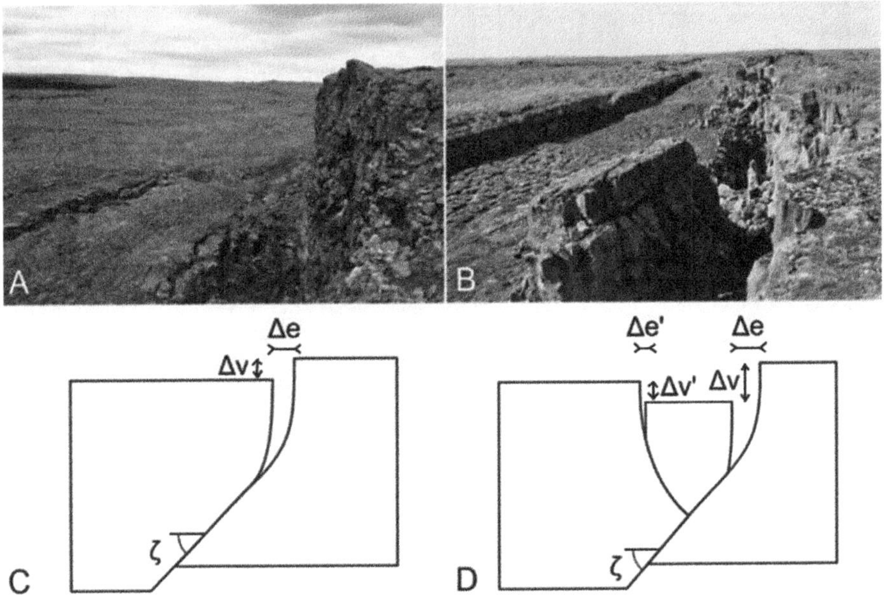

Figure 2.19. *Relationships between open fractures geometry at the surface and normal shear at depth (modified from Angelier* et al. *1997) (photos A and B: Françoise Bergerat©)*

COMMENT ON FIGURE 2.19.– *Examples of a single scarp fault (A) and two faults forming a small asymmetrical graben (B) in the Mófell fissure swarm (NVZ). Geometrical relations between offset, fracture dilation and dip of the fault plane in the case of a single fault (C) or two antithetical faults (D). Δv: scarp height; Δe: fracture dilation; ζ: dip of the fault plane at depth.*

The relations between the fracture dilation and scarp height measured at the surface, on the one hand, and the actual dip of the fault plane at depth on the other hand (Figure 2.19(C) and D) are elementary for the model with a simple escarpment:

$$\tan \zeta = \Delta v / \Delta e \qquad [2.1]$$

and slightly more complicated for the model with an asymmetrical gap:

$$\tan \zeta = (\Delta v - \Delta v')/(\Delta e + \Delta e') \qquad [2.2]$$

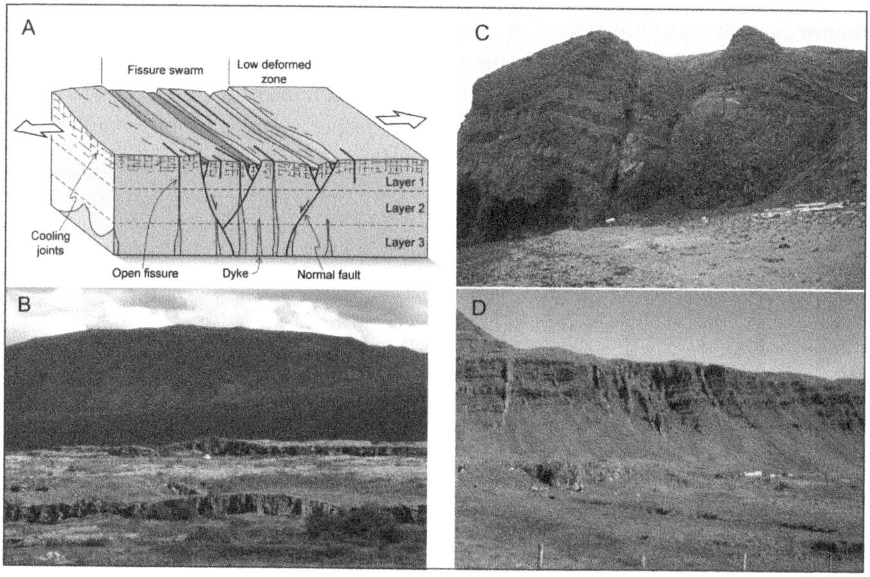

Figure 2.20. *Vertical partitioning of the deformation in fissure swarms*

COMMENT ON FIGURE 2.20.– *(A) Block diagram showing the different structures as a function of depth (modified from Dauteuil* et al. *2001). The three levels of deformation are illustrated by (B) recent faults affecting postglacial lavas, in the Þingvellir fissure swarm, in the WVZ (layer 1), (C) normal faults in basaltic rocks of Upper Miocene-Lower Pliocene (3.3–8.5 My) in the Tjörnes Peninsula, on the north coast of Iceland (layer 2) and (D) dykes in basaltic rocks of the Upper Miocene (8.5–10 My) in the Berufjörður, on the east coast of Iceland (layer 3) (photos B–D: Françoise Bergerat©).*

The dip of the faults reconstructed by this way is of the order of 60°–72° (Angelier *et al.* 1997; Sonnette *et al.* 2010). Comparison with the dip data of older faults observed on the eroded shoulders of the rift (Guðmundsson *et al.* 1996; Bergerat and Angelier 1998), corresponding to paleo-depth of 1–2 km, indicates that this determination is valid. On the other hand, these calculations do not allow for determining the exact depth at which the change in slope of the fault plane occurs. However, it is likely that the horizontal tension dominates in the first few hundred meters due to the free surface and low lithostatic pressure, hence the development of open cracks (Figures 2.20(A) and (B)). Normal shear plays a fundamental role at an intermediate depth (probably of the order of one kilometer) with higher lithostatic pressure and normal faults with a mean dip of 65°–75°. Finally, at greater depths, the fluid pressure is high enough to allow the injection of magma (Dauteuil *et al.* 2001). Erosion in the oldest lavas allows the deeper parts to be observed: normal shear planes and dykes (Figures 2.20 (C) and (D)).

Figure 2.21. *Normal faults and dykes in tertiary lava flows in Glaðheimar, Borgarnes region, western Iceland, view toward the north (Françoise Bergerat©)*

This structuring of the rift deeply marks the Icelandic landscape with its superimposed lava flows, sometimes slightly inclined and cut by normal faults and dykes, ancient witnesses, moved away from the rift axis over time and exposed thanks to erosion (Figure 2.21).

Although it is accepted that deep magma has a major effect on how the tectonic structures develop in the Icelandic crust, the way normal faults develop is an important point still debated. A number of models favor an upward propagation from the depth (Grant and Kattenhorn 2004; Tentler 2005), while others favor a downward one from the surface (Acocella *et al.* 2003).

For Grant and Kattenhorn (2004), the high tensile stresses at the upper extremity of a deep fault are large enough to exceed the lithostatic compression and tensile strength of the rock, which can lead to the growth of a vertical fracture at the fault end, with a narrow monocline developing above it (Figures 2.22 (A1) and (A2)). Once the fracture reaches the surface, vertical displacement and opening occur along the fault scarp (Figure 2.22(A3)). The depth of the upper extremity of the normal shears is estimated, from their models, to be between 250 and 500 m.

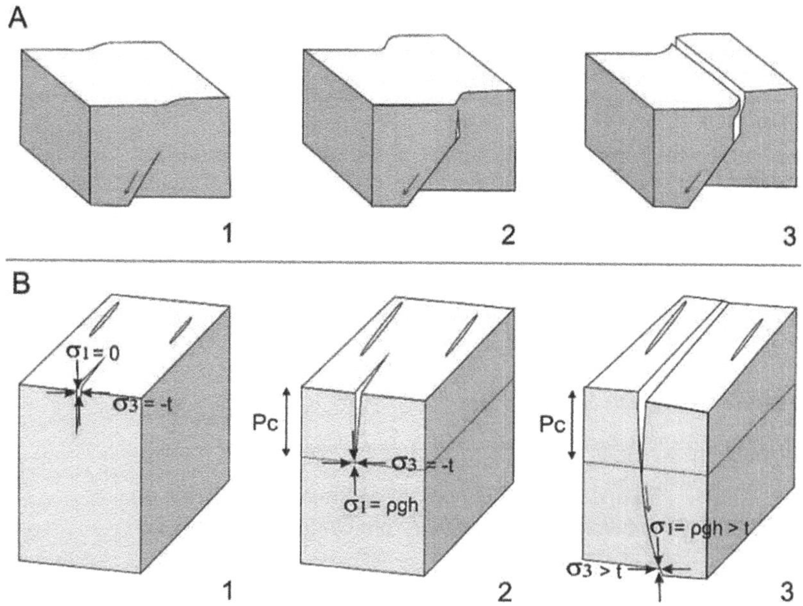

Figure 2.22. *Simplified models of the formation and growth of a normal rift fault*

COMMENT ON FIGURE 2.22.– *(A) Upward propagation (modified from Grant and Kattenhorn 2004). A normal fault initiates at depth and propagates upwards (1); then a vertical fissure forms at its upper extremity and a flexure develops at the surface (2); finally the fissure reaches the surface and cuts the flexure, creating a fault scarp (3). (B) Downward propagation (modified from Acocella* et al. *2003). A vertical open fissure is created at the surface (1), then develops to a critical depth Pc (2) and finally transforms into a normal fault by continuing to grow toward depth (3). Tensile strength: t; rock density: ρ; acceleration of gravity: g; depth: h.*

For Acocella *et al.* (2003), at the beginning of fracturing, a point on the surface is subject to a zero vertical stress σ_1[2] (lithostatic) and to a horizontal stress (tectonic) $\sigma_3 > -t$ (Figure 2.22(B1)). If this latter increases, the stress at the considered point reaches the tensile strength, and a tension fracture is created at the surface. Once formed, it propagates at depth (Figure 2.22(B2)) until σ_1 and σ_3 exceed the tensile strength of the rock; normal shear then develops (Figure 2.22(B3)). Depending on the

2 The notations σ_1, σ_2 and σ_3 correspond respectively to the maximum, intermediate and minimum principal stress axes of the stress tensor. The notation σ_{Hmax} corresponds to the maximum horizontal stress, *i.e.* σ_1 for a strike-slip regime and σ_2 for a normal regime; in both cases the direction of extension is given by σ_3.

failure criterion adopted (Mohr-Coulomb or Griffith), this critical depth would be between 260 and 780 m.

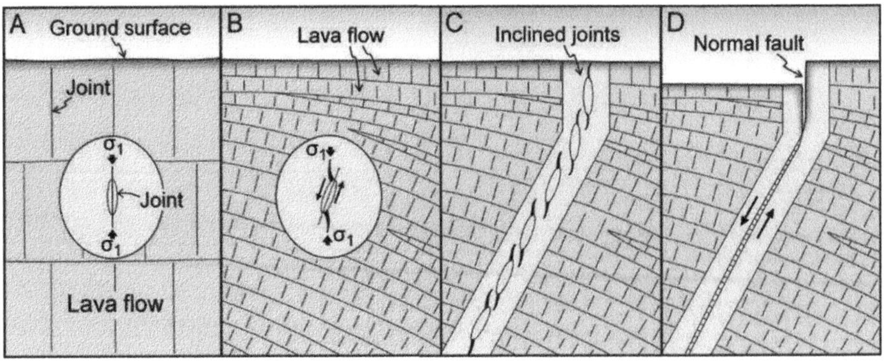

Figure 2.23. *Development of a normal fault from en échelon joints (modified from Guðmundsson 1992)*

COMMENT ON FIGURE 2.23.– *Initially, the cooling joints in the basaltic layers are vertical and parallel to σ_1 (A); they become oblique during the tilting of the lava pile, representing then potential shear fractures (B); finally, the latter can meet each other (C) and give rise to a true normal fault (D).*

Finally, other authors advocate propagation both downward and upward (Guðmundsson 1992; Martel and Langley 2006). Guðmundsson (1992) thus suggested two initiation mechanisms: (i) a nucleation from tension fractures created at the surface and propagating at significant crustal depths, and (ii) a nucleation on sets of *en échelon* joints during the tilting of lava flows (Figure 2.23), both mechanisms occurring simultaneously. Based on the tensile strength of the rock, Guðmundsson (1992) determines the critical depth at which the joints begin to join each other to initiate a normal fault and proposes 0.5–1.5 km, which is consistent with field observations made in eroded areas such as coastal cliffs.

In detail, since the fracture fields are regularly overlapped by new basaltic lava flows, the development of fractures in the youngest ones is guided by the presence of underlying pre-existing fractures (MacDonald 1957; Duffield 1975; Peacock and Parfitt 2002; Martel *et al.* 2006; Sonnette *et al.* 2010).

Numerous studies have concerned fissure swarms, suggesting or refuting close links between faults and open fractures at the surface of active zones, on the one hand, and deep dyke swarms, on the other hand (Helgason and Zentilli 1985;

Forslund and Guðmundsson 1991; Guðmundsson 1995a, 1995b, 2003; Tentler 2005; Paquet *et al.* 2007). In other contexts, such as Afar, some authors (Rowland *et al.* 2007) assume that (i) normal faults are created or reactivated above the propagating dykes and (ii) pre-existing subvertical cooling joints are activated as open fractures at the upper extremity of normal faults. However, in Icelandic paleo-rift zones, the number of dykes largely exceeds the number of normal faults in the same sections (Forslund and Guðmundsson 1991; Guðmundsson 2003; Paquet *et al.* 2007), a feature which would tend to invalidate this hypothesis.

The way the fault segments merge and the changes in fault architecture produced by this coalescence have been particularly well documented in different rock types and tectonic contexts. Numerous works have analyzed fault configurations (Peacock 1991; Cartwright *et al.* 1995; Childs *et al.* 1995; Huggins *et al.* 1995), in the horizontal plan or the vertical one (Peacock and Zhang 1993; Childs *et al.* 1996; Mansfield and Cartwright 1996; Schöpfer *et al.* 2006), demonstrating that connections between different parts of normal faults occur both horizontally and vertically, and this over a wide range of scales. The Icelandic rift is a remarkable natural laboratory for understanding the mechanisms underlying fault growth.

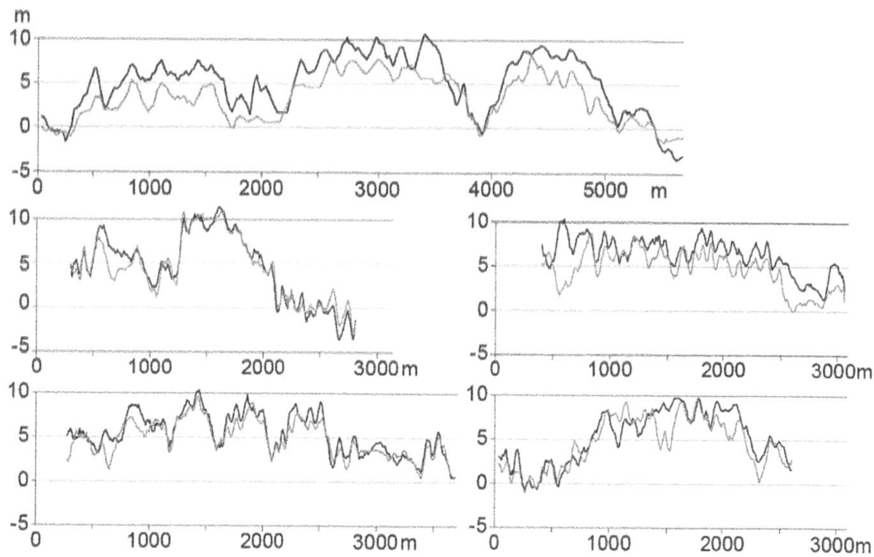

Figure 2.24. *Along-strike profiles of vertical offsets for some faults in the Vogar fissure swarm (modified from Villemin and Bergerat 2013). Profiles established at distances of 20 m (in black) and 40 m (in gray) away from the main discontinuity*

Establishing longitudinal profiles of the offsets along the normal faults bordering or within the fissure swarms allows to establish profiles of offset values along these faults (Figure 2.24) and to propose linkage and propagation models (Villemin and Bergerat 2013). These profiles show overall that the offsets are greater in the central part of the fault segments and decrease at their extremities, but also that the large faults are, in fact, made up of several connected segments.

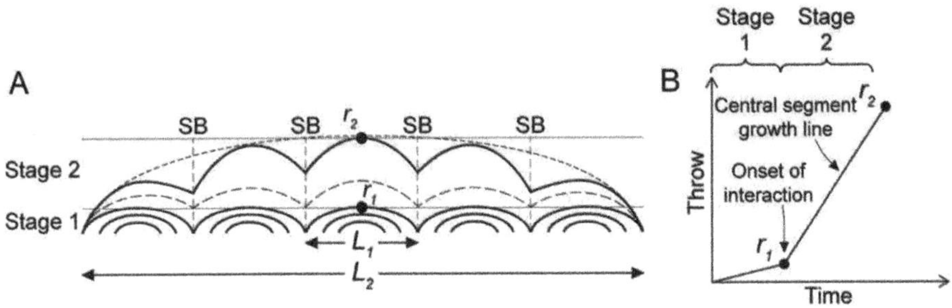

Figure 2.25. *Conceptual model of normal fault growth in the Icelandic rift (from Villemin and Bergerat 2013; adapted from Cowie and Roberts 2001; Roberts and Michetti 2004; Papanikolaou and Roberts 2007)*

COMMENT ON FIGURE 2.25.– *(A) Length-offset profile model. At the end of stage 1, the linkage occurred between the different fault segments and they interacted over large distances. (B) Graph illustrating the offset growth as a function of time for the central segment shown in A. SB: segment boundary; r_1: maximum offset on each segment prior to linkage; r_2: maximum offset on the entire fault after linkage; L_1: length of a single segment; L_2: length of the fault after linkage.*

The detailed analysis of these profiles (Villemin and Bergerat 2013) allows us to establish a growth model for normal rift faults (Figure 2.25). The first stage corresponds to the isolated growth of each fault segment, during which offset grows moderately with respect to length; in a second stage, when the different segments are connected, offset increases while there is no (or little) growth in length. In detail, in stage 2, there is first a period during which the offset grows in the central part of the segments but remains low near the junction areas, and then a period during which the offset growth at the connection points makes it possible to make up for the "deficit".

2.3.2. *The transform zones*

The presence of transform zones in Iceland is intimately linked to the existence of rift jumps (section 2.3.1.1). The last rift jump in the north of the island dates from about 8–8.5 My and was completed about 3 My ago (Garcia *et al.* 2003). In the south, the last rift jump started 2–3 My ago (Jóhannesson *et al.* 1990) and is still ongoing. Consequently, the two transform zones do not have the same stage of development and the geometric configuration of the faults that constitute them reflects this different degree of maturity. Within each of these two transform zones, the fault zones may present different degrees of development. It is thus possible to observe different systems from zones of diffuse deformation without major surface faults to zones where the faults are well individualized and in the same direction as the direction of the plate motions.

2.3.2.1. *The Tjörnes Fracture Zone*

In the north of the island, the TFZ connects the NVZ to the KR. We have seen that seismic activity defines a WNW-ESE zone, 120 km long and 70–80 km wide, composed of three approximately parallel seismic lineaments which are, from south to north, the DL, the HFF and the GL (Figure 2.26; see also section 2.2.1.2). Two of these are partly observable on land (Sæmundsson 1974; Björnsson 1985) and microseismic data have also been used to characterize the epicenters and fault planes of large historical earthquakes (section 2.2.1.2) (Stefánsson *et al.* 2008). The general movement in the WNW-ESE direction is right-lateral.

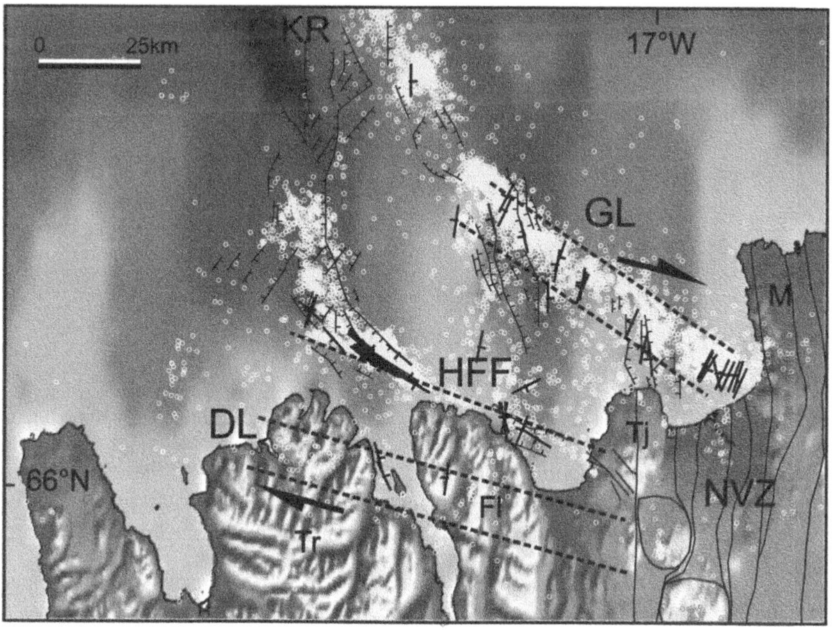

Figure 2.26. *The Tjörnes Fracture Zone*

COMMENT ON FIGURE 2.26.– *Thin solid lines: Faults mapped onshore or detected by seismic reflection and fissure swarm boundaries in the NVZ; thick solid lines: major seismic faults; dashed lines: Dalvík (DL) and Grímsey (GL) lineaments and the Húsavík-Flatey Fault (HFF); white dots: epicenters of the 1994–1998 earthquakes. Fl, M, Tj and Tr: Peninsulas of Flateyjarskagi, Melrakkaslétta, Tjörnes and Tröllaskagi (faults and earthquakes according to MacMaster 1977; Rögnvaldsson* et al. *1998; Dauteuil* et al. *2002).*

2.3.2.1.1. Lineaments of the TFZ

The DL, currently the least seismic one, is expressed on land only by morphological traces such as the direction of the coast of the Tröllaskagi Peninsula east of Dalvík, the Dalsmynni valley in the Flateyjarskagi Peninsula (Fjäder *et al.* 1994) and a few NNE-SSW to NNW-SSE shear corridors (respectively, sinistral and dextral) a few meters wide, often associated with hydraulic breccias and fracture schistosity (Figure 2.27) (Bergerat *et al.* 1992; Bergerat and Angelier 1999). There is no evidence of a major fault in the transform direction (Långbacka and Guðmundsson 1995). The few seismic faults in this area are submeridian and sinistral (Rögnvaldsson *et al.* 1998).

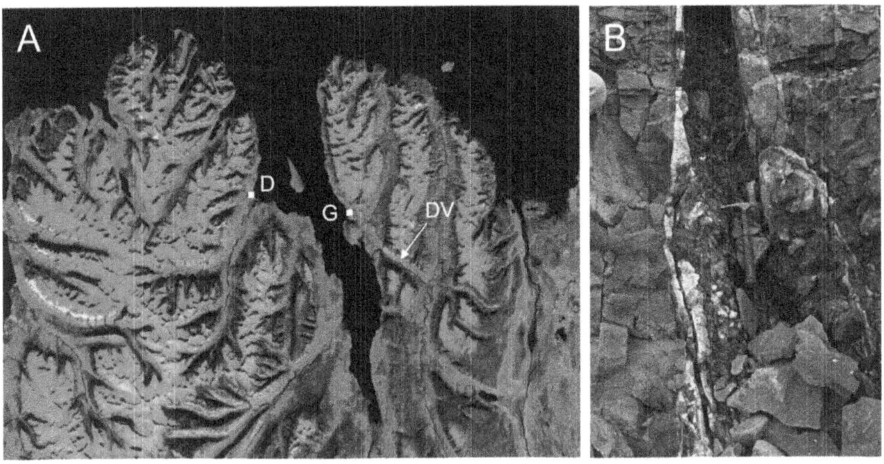

Figure 2.27. *The Dalvik Line. (A) Morphostuctural traces in the Flateyjarskagi and Tröllaskagi Peninsulas (SPOT 4 IR image). (B) Submeridian strike-slip fault with crushed zone within the Tertiary basalts, on the coast, north of Grenivík (Françoise Bergerat©). D: Dalvík; G: Grenivík; DV: Dalsmynni Valley*

The GL has been active since about 2 My; it is currently the northern limit of the TFZ and is located entirely offshore. The seismicity emphasizes a general WNW-ESE direction, but the focal mechanisms indicate left-lateral motions on NNE-SSW to N-S faults (Rögnvaldsson *et al.* 1998; see also section 2.2.1.2). The main structures (defined in seismic reflection) associated with it are normal faults and grabens, 5–20 km wide, oriented NNW-SSE and parallel to those of the northern part of the Fremri Námar fissure swarm (NVZ) (Guðmundsson and Bäckström 1991; Guðmundsson *et al.* 1993). No major structures are visible on land, although seismites in post-glacial sediments in the Melrakkarslétta Peninsula testify to its activity (Figure 2.28).

The HFF is the most important structure in the TFZ; it is observable on land in the Tjörnes and Flateyjarskagi Peninsulas. It is marked offshore by a valley 5–10 km wide and 3–4 km deep. Its vertical offset can exceed 1 km (Tryggvason 1973).

Figure 2.28. *Cosismic loading deformations in Holocene glacio-marine sediments (ca. 11.0 Ky) in Kópasker, west coast of the Melrakkarslétta Peninsula, at the eastern end of the Grimsey Line. Height 1.20 m (Brigitte Van Vliet-Lanoë©)*

In the Tjörnes Peninsula, the HFF can be tracked over about 25 km between the NVZ and the coast, bringing Tertiary lavas to the north into contact with Upper Pleistocene lavas to the south (Sæmundsson 1974). In detail, the HFF consists of several segments, parallel and/or *en échelon*, affecting rocks of various natures and ages (Figure 2.29), and showing both normal and strike-slip motions in accordance with a general dextral transtensive movement.

The stress inversion of fault slip data, as well as that of the focal mechanisms of earthquakes, indicates a direction of σ_3 partitioned into two stress states: an extension perpendicular to the HFF (N 50° E) and a strike-slip motion in the transform direction (N 130° E) (Garcia *et al.* 2002). Near Húsavík, the vertical displacement on the HFF is more than 200 m and can reach 1,400 m (Guðmundsson *et al.* 1993).

The general arrangement of the main faults in the Tjörnes Peninsula has been determined from aerial and spatial imagery (Mamula and Voight 1985; Garcia and Dhont 2005) (Figure 2.30) and from geological mapping (Sæmundsson 1974; Guðmundsson *et al.* 1993; Jóhanesson and Sæmundsson 1998).

Figure 2.29. *The Húsavík-Flatey Fault, on the coastal cliff, under the former garbage dump north of Húsavík. The HFF brings into contact the Middle Pleistocene tills (right) and the Early Pleistocene lithified tillites (left). The orange-colored terrains in the faulted zone are Eemian sands (see detailed section in Van Vliet-Lanoë* et al. *2005) (Brigitte Van Vliet-Lanoë©)*

Transtension is attested by the existence of pull-apart structures such as the Botnsvatn and Höskuldsvatn lakes (Figures 2.30 and 2.31).

Toward the east, the HFF segments connect to the Þeistareykir fissure swarm (NVZ) either by a progressive curvature of the HFF faults or by the imbrication of these faults with those of the N-S rift (Figure 2.30) (Guðmundsson *et al.* 1993; Garcia and Dhont 2005).

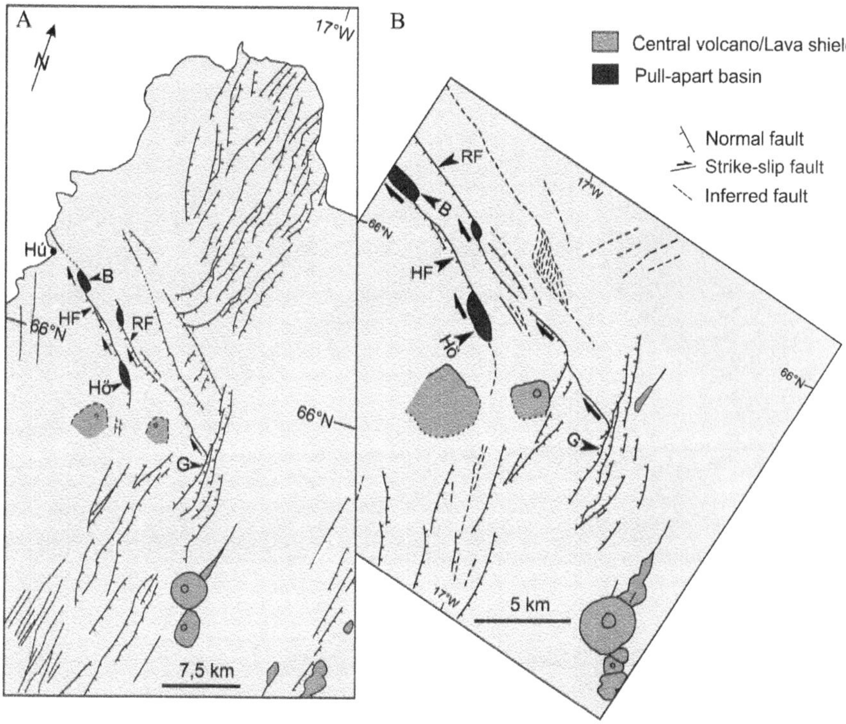

Figure 2.30. *Geometry of the TFZ and junction with the Þeistareykir fissure swarm (NVZ) (modified from Garcia and Dhont 2005)*

COMMENT ON FIGURE 2.30.– *(A) Interpretation of the ERS-1 radar image from 18/09/1992. (B) Interpretation of the SPOT panchromatic image from 14/07/1990 (reference KJ 717-213). B: Botnsvatn Lake; G: Guðfinnugjá Fault; HF: Húsavík Fault; RF: Reyðará Fault; Hö: Höskuldsvatn Lake; Hú: Húsavík.*

In the northern part of the Flateyjarskagi Peninsula, the HFF is expressed by a 3–5 km wide, intensely deformed complex zone. It includes numerous strike-slip, normal and oblique-slip faults, mineralized veins with calcite, quartz and zeolites, dykes and crushed zones (Figure 2.32) (Young *et al.* 1985; Guðmundsson 1993, 2007; Fjäder *et al.* 1994; Guðmundsson and Fjäder 1995; Jancin *et al.* 1995; Bergerat *et al.* 2000; Garcia 2003). The major faults are broadly NNE-WSW in the southern and central parts of the peninsula and become NW-SE in the north. In the highly deformed zone, they may constitute corridors of completely crushed rock, sometimes several meters wide. The directions and dips of the lava layers also change toward the north of the peninsula.

Figure 2.31. *The Húsavík-Flatey Fault (HFF) in the Tjörnes Peninsula. (A) Aerial photography. (B) East-southeast view of Lake Botnsvatn (Bo) showing a pull-apart structure located on the HFF (Águst Guðmundsson©, London). (C) Same photograph with indication of the two blocks separated by the fault; the block on the right is shown in transparency and the dextral movement of the fault is indicated*

The most interesting – and most debated – structures are the dykes. According to Young *et al.* (1985), these have an N-S to NNE-SSW direction in the southern and central parts of the peninsula and become ENE-WSW to WNW-ESE north of a Gil-Látur Line (Figures 2.33(A) and (B)). Even if the directions measured by Fjäder *et al.* (1994) are practically the same (N-S to NNE-SSW in the south and E-W to NW-SE in the north), the controversy concerns the interpretation of these different directions.

Figure 2.32. *Types of tectonic deformation encountered on the northern coast of the Flateyjarskagi Peninsula. (A) Crushed and altered zone (Sebastian Garcia©). The edges of the crushed zone are underlined by dashed lines. The basalts are also intensely fractured near these edges. (B) Normal faults. (C) Mineralized veins (photos B and C: Françoise Bergerat©)*

For Young *et al.* (1985), the clockwise rotation of faults and dykes in the north of the peninsula results from block rotations within a shear zone 11 km wide and limited by the Gil-Látur line and the HFF.

In contrast, for Fjäder *et al.* (1994), the intense deformations are concentrated in a 3–5 km wide zone along the northern coast of the peninsula and the changes in lava direction are original and do not result from a later tectonic rotation. These authors also postulate the existence of two distinct dyke swarms based on a difference in their thickness. Subsequent discussions (Guðmundsson and Fjäder 1995; Jancin *et al.* 1995) have focused on the interpretation of these dyke swarms.

Subsequent works in the Flateykarskagi and Tröllaskagi Peninsulas (Bergerat *et al.* 2000; Garcia 2003) have complemented these small-scale observations with an analysis of striated faults and reconstruction of the different stress states controlling the deformations (Figures 2.33(C) and (D)). For Bergerat *et al.* (2000), the complexity of this domain, both in terms of stresses and structures, cannot be explained by a single process. These authors distinguish two sets of dykes: (i) a set sub-parallel to the HFF (E-W to ESE-WNW, in red in Figure 2.33(A)), existing only in the northern part of the peninsula, which would be consistent with the state of stress in tension related to ridge migration suggested by Guðmundsson (1993, 1995a); and (ii) a N-S dyke set in the south and center of the peninsula and having a general clockwise curvature of about 50° in the north, instead of the 110° given by Young *et al.* (1985) for which there is only one family of dykes. This curvature is related to a dextral shear zone.

Based on new paleomagnetic investigations, Young *et al.* (2020) have strengthened their hypothesis of significant rotations occurring in a 20 km wide zone.

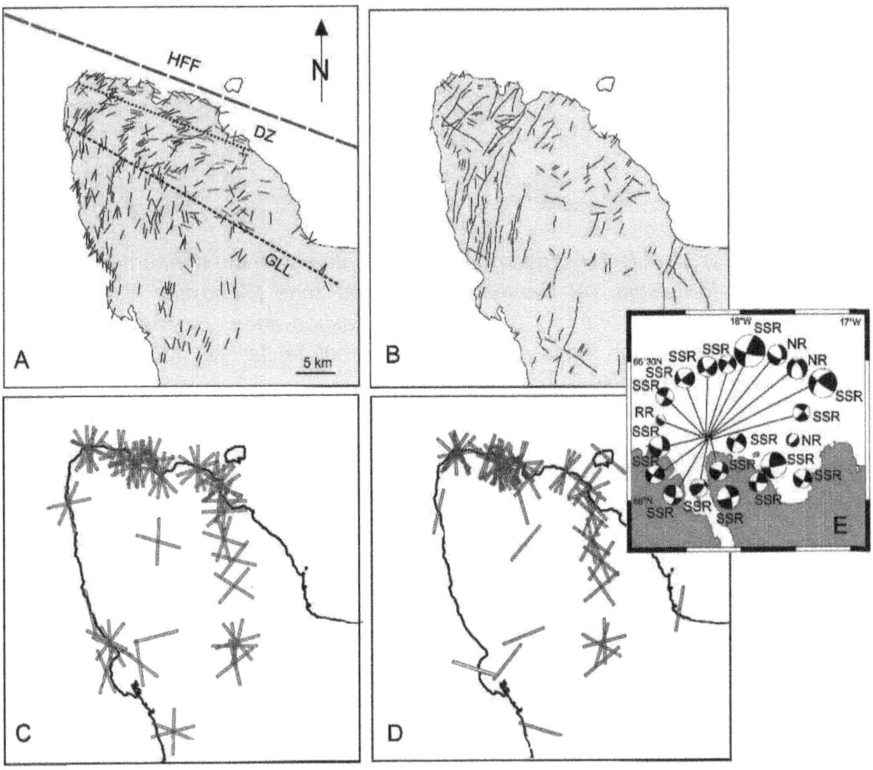

Figure 2.33. *Neogene and present-day faults and stresses in the Flateyjarskagi Peninsula*

COMMENT ON FIGURE 2.33.– (*A and B) Directions of dykes (A) and normal faults (B) (modified from Young* et al. *1985). (C) Directions of horizontal maximum principal* *stress (σ_{Hmax}) calculated by inversion of fault slip data for the populations of strike-slip faults (C) and normal faults (D) (modified from Homberg* et al. *2010). (E) Selection of earthquake focal mechanisms M > 2 of the different regimes, for the period 1995 to 1997 (size of circles proportional to the magnitude) (modified from (Garcia* et al. *2002)). SSR: Strike-slip regime; NR: Normal regime; RR: Reverse regime; HFF: Húsavík-Flatey Fault. The southern limit of the intense deformation zone (DZ) of Fjäder* et al. *(1994) and the Gil-Látur Line (GLL) of Young* et al. *(1985) are also shown in (A).*

It should be noted that the lack of certainty regarding the chronological relationships between the various structures could indicate that there is not a well-defined succession of different processes, but rather alternating ones, which can be very short – in the order of a few years – as suggested by the variety of focal mechanisms for present-day earthquakes (Figure 2.33E) (Garcia *et al.* 2002).

2.3.2.1.2. Deformation partitioning in the TFZ

The inversion of fault data collected in the Tröllaskagi and Flateyjarskagi peninsulas led to the reconstruction of four stress regimes including both normal and strike-slip modes (Figure 2.34). The two main regimes are characterized by E-W and NE-SW directions of extension (σ_3), consistent with the right-lateral movement of the HFF. The other two, called opposite, are inconsistent with this motion and result from stress permutations. The relations between these different stress regimes and modes involve not only permutations σ_1/σ_2 and σ_2/σ_3, but also σ_1/σ_3. Each main regime is, ultimately, characterized by the angular relationship between σ_3 and the transform direction (Figure 2.34). This angle is able to vary between 20°–30° and 80°–90° and reflects the changes of the friction coefficient along the HFF. Analysis of the earthquake focal mechanisms (Angelier *et al.* 2000; Garcia 2003) shows that such variations, ranging from moderate to very weak coupling on a low-friction coefficient fault (weak fault), can occur very rapidly.

Single and multi-fault numerical models have been carried out to simulate the distribution of slips and stresses in the TFZ (Homberg *et al.* 2010). They show that the current width of the TFZ does not correspond to a permanent deformation distribution in this wide zone, implying synchronous activity on all major faults, but rather to a transfer of slip from one major fault to another over time, with previous slip zones being locked or not very active. The predominance of one fault does not exclude however that moderate activity persists along the other structures. The interactions between faults explain the modifications in the stress trajectories that occur near – but also sometimes quite far away – from the active zones.

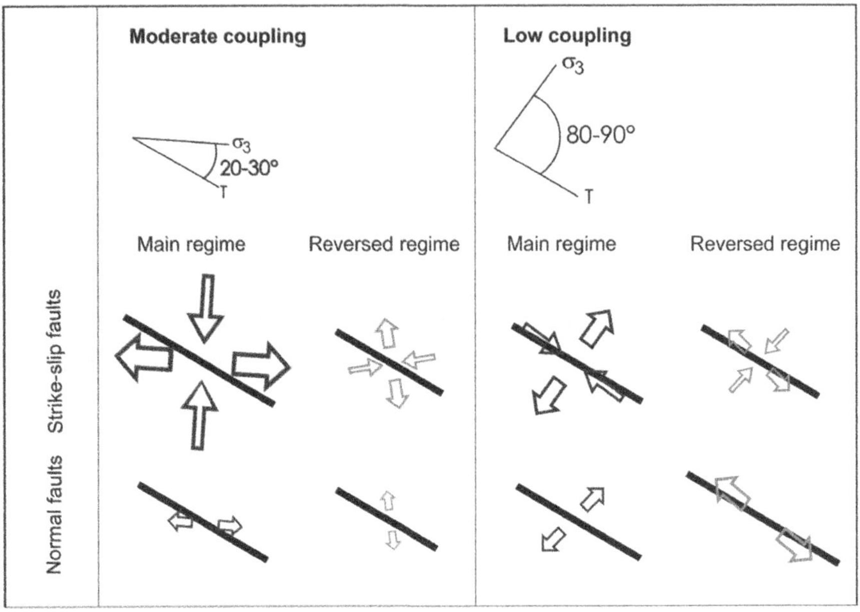

Figure 2.34. *Stress states in the Flateyjarskagi Peninsula (from Angelier* et al. *2000; Bergerat* et al. *2000) characterized from fault slip data*

COMMENT ON FIGURE 2.34.– *Each major stress state includes a main regime and an opposite regime. The diverging arrows indicate the average direction of extension (σ_3) for the strike-slip and normal regimes, the converging arrows indicate the direction of compression (σ_1) for the strike-slip regimes. The size of the arrows increases with the importance of the stress states considered. The thick black line represents the general direction of the transform zone. The angle between the transform direction (T) and the average direction of extension (σ_3) is shown for the two main regimes.*

Among the different models tested, it appears that those simulating large slips along the DL and GL do not correctly reproduce the orientation of past and present stresses, reconstructed or measured in the TFZ. The HFF seems to have accommodated most of the movement so far, making it the major transform structure. However, it should be noted that due to the northward propagation of the NVZ (Guðmundsson *et al.* 1993), the GL has a strong potential to become the main fault in the TFZ. Consequently, although there is no evidence that the activity of the present-day main fault of the TFZ (the HFF) is decreasing, it is likely that the main transform movement will be transferred to the GL in the future.

2.3.2.2. *The South Iceland Seismic Zone*

Considering the spatial distribution of earthquakes, the SISZ is an east-west transform zone, 70–80 km long and 20–30 km wide, which connects the WVZ to the EVZ (Stefánsson *et al.* 1993; Böðvarsson *et al.* 1996). In fact, it is located at the junction of three rift segments (Figure 2.35): the RP, which is the onshore extension of the RR, the WVZ and the EVZ. The EVZ is the youngest part of the Icelandic rift and has been active since 2–3 My (Jóhannesson *et al.* 1990).

2.3.2.2.1. Structures of the SISZ

While the geometry of the rift segments resembles that of an overlapping spreading center (OSC) more than that of a transform structure, the seismicity of the SISZ is, nevertheless, in agreement with a left-lateral transform movement between the EVZ and RP. Submeridian faults, arranged side by side and dextral, mainly accommodate the deformation (Einarsson and Eiriksson 1982; Einarsson *et al.* 2005). They are characterized by (i) seismological data for the present (Rögnvaldsson and Slunga 1994; see also section 2.2.1.2), (ii) historical seismicity data for the Holocene (Bjarnasson *et al.* 1993), and more generally by (iii) geological mapping (Sæmundsson and Einarsson 1980; Jóhannesson *et al.* 1990) and (iv) structural studies (Passerini *et al.* 1990; Guðmundsson and Brynjolfsson 1993; Guðmundsson 1995b; Bergerat *et al.* 1998, 1999; Bergerat and Angelier 2000).

However, looking at the Pleistocene faults configuration, we notice that the width of the SISZ is, in fact, of the order of 60 km and thus larger than that defined by microseismicity (Guðmundsson 1995b; Van Vliet-Lanoë *et al.* 2020a). At the regional scale, the major faults have directions varying from N-S to ENE-WSW.

Despite some heterogeneity (Guðmundsson 1995b; Bergerat *et al.* 1998), the directions N 30°–40° E correspond mainly to normal faults or tension fractures, while the directions N 0°–20° E characterize dextral strike-slip faults and the directions N 60°–70° E reflect left-lateral movements.

The Vörðufell massif (Figure 2.36), located on the northern edge of the SISZ, represents an example of this complex geometry (Bergerat *et al.* 1998; Bergerat and Angelier 2000). The major fractures (faults and dykes) are N-S and ENE-WSW (Figure 2.36(B) and (C)). In detail, the NE-SW to ENE-WSW striated fault planes frequently show traces of two different motions (normal and strike-slip) (Figure 2.36(D)).

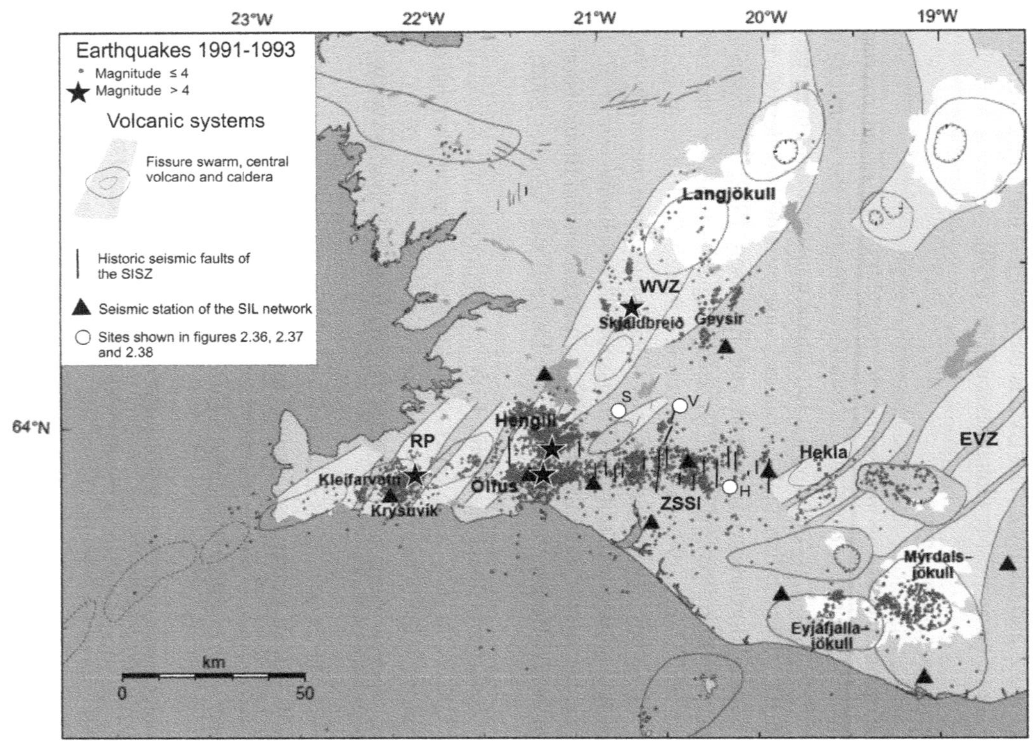

Figure 2.35. *The South Iceland Seismic Zone (SISZ) between the East and West Volcanic Zones (EVZ and WVZ) and the Reykjanes Peninsula (RP) (modified from Jakobsdóttir* et al. *2002). H: Heiðarbrekka; S: Seyðishólar; V: Vörðufell*

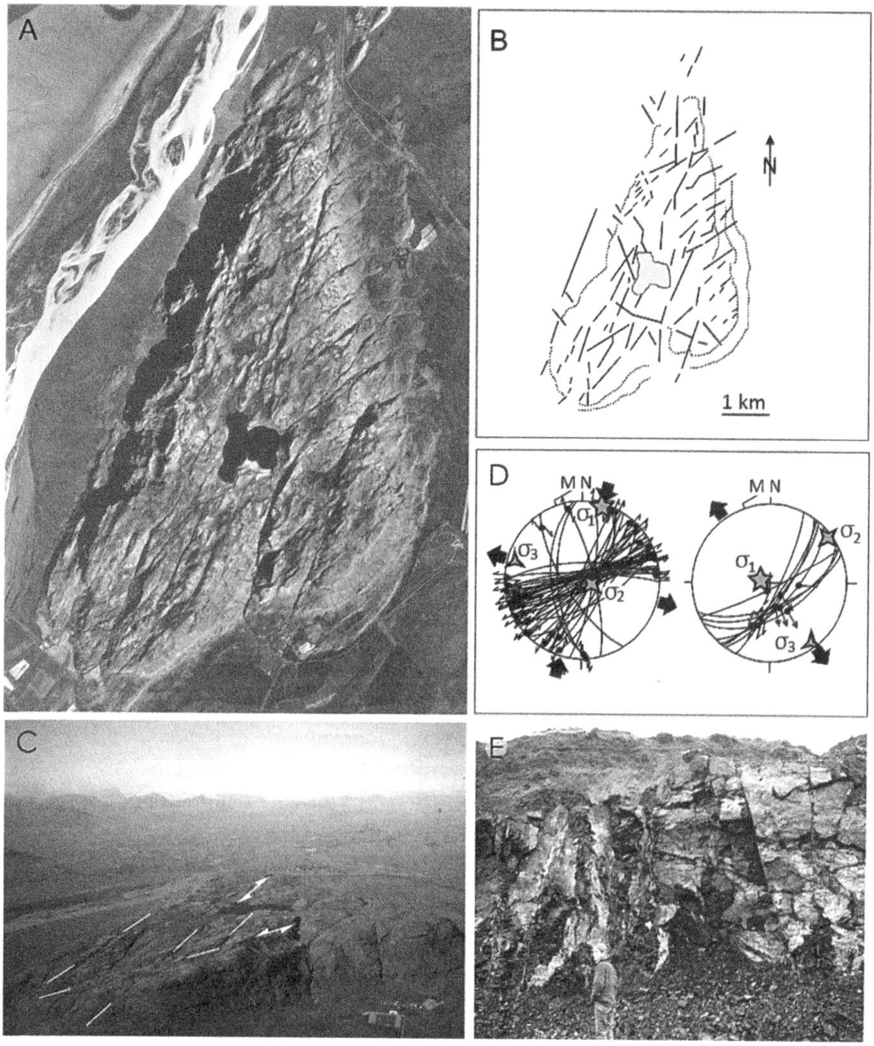

Figure 2.36. *The Vörðufell massif (SISZ)*

COMMENT ON FIGURE 2.36.– *(A) Aerial photograph. (B) Map of the faults affecting the massif, established from aerial photography (from Bergerat and Angelier 2000). Solid lines: set of major fractures, dotted lines: cliffs. (C) Oblique aerial view (Águst Guðmundsson©, London). Double arrows: Direction of the strike-slip movements on the major faults N-S and ENE-WSW (highlighted by white lines). (D) Cyclographic representation of the measured strike-slip and normal faults (planes and striations) and of the axes of the calculated stress tensor (Schmidt's projection, lower hemisphere). The large black arrows indicate the directions of compression and extension. (E) Set of strike-slip faults in a small quarry in the north of Vörðufell (Françoise Bergerat©).*

Figure 2.37. *Figures of water escape (cosismic deformation) in the Rangá interglacial formation (around 125 Ky), in Heiðarbrekka, near the Ytri-Rangá river (Brigitte Van Vliet-Lanoë©)*

Evidence for the permanent functioning of the SISZ since the Pleistocene to the Present also exist in the Holocene formations. Figure 2.37 shows cosismic deformations in the Rangá formation, at MIS 5 (Marine Isotope Stage) in the eastern part of the SISZ (Van-Vliet Lanoë *et al.* 2020a). These deformations are located to the south of the traces of the Minivellir fault (1630 earthquake; see section 2.3.2.2), thus showing the reutilization of the same accidents during successive seismic crises over time.

At the other end of the SISZ, near its junction with the WVZ , the example of the Seyðishólar volcanic cone, crossed by normal-strike-slip faults (Figure 2.38), also illustrates the recent deformation of the southern part of the island. These oblique motions mark the transition between the purely strike-slip movements of the SISZ and the normal rift-related movements.

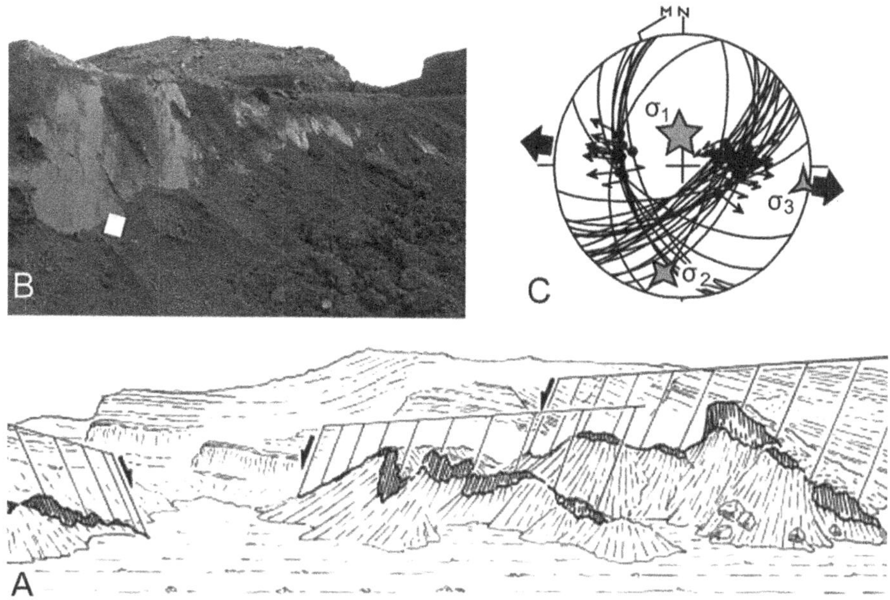

Figure 2.38. *Normal faults in Holocene pumice, postglacial Seyðishólar crater (modified from Bergerat and Angelier 2000)*

COMMENT ON FIGURE 2.38.– *(A) View toward the south-southwest of the faults delimiting a small graben . (B) A mirror of a submeridian fault with oblique striation and normal-dextral movement (Françoise Bergerat©). (C) Cyclographic representation of the total fault population (planes and striations) and of the axes of the calculated stress tensor (Schmidt's projection, lower hemisphere). The large black arrows indicate the direction of extension.*

2.3.2.2.2. Historical seismic faults

Historical data on earthquakes in the SISZ date back to 1150 for the oldest ones. From 1700 onwards, the descriptions are sufficiently complete to allow their magnitudes to be estimated with good precision for those above M 6 (Stefánsson *et al.* 1993). The location of surface traces of major prehistoric and historical earthquakes in the SISZ is thus well known from field work and study of written sources (Thoroddsen 1899, 1905; Einarsson *et al.* 1981; Einarsson 1991). General maps drawn up by Einarsson *et al.* (2005) show a general *en échelon* pattern of major dextral faults, with a spacing of 1–6 km (Einarsson *et al.* 2005; see also

Figure 2.39). Their length, calculated using scaling laws, ranges from 10 to 35 km (Roth 2004). These faults are submeridian, i.e. subperpendicular to the general direction of the SISZ. However, conjugate sinistral segments of lesser length are frequently present (Bjarnason *et al.* 1993; Bergerat and Angelier 2001; Angelier and Bergerat 2002; Bergerat and Angelier 2003; Bellou *et al.* 2005; Clifton and Einarsson 2005; Bellou 2006; Guðmundsson 2007; Bergerat *et al.* 2011).

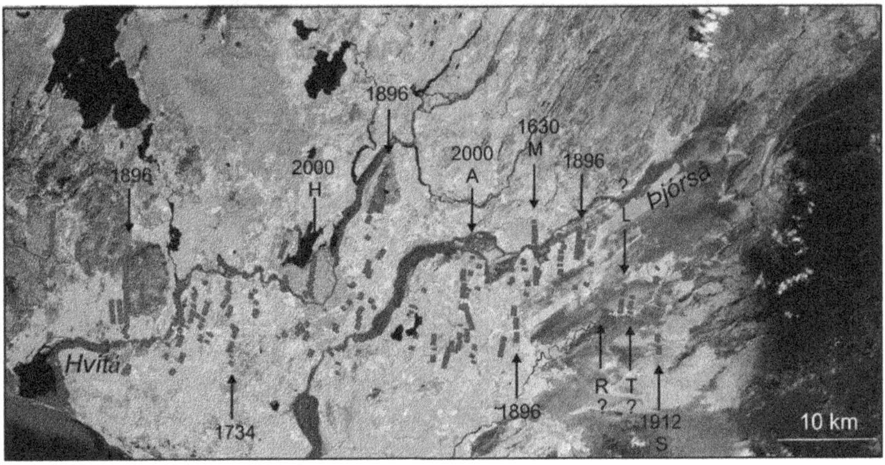

Figure 2.39. *Traces of historical and present major seismic faults in the South Iceland Seismic Zone (from Einarsson and Eiríksson 1982; Einarsson* et al. *2005), plotted on a Landsat image of the SISZ. H: Hestfjall fault (2000); Á: Árnes fault; L: Leirubakki fault; M: Minnivellir fault; R: Réttarnes fault; T: Tjörvafit fault; S: Selsund fault*

Several of these historical seismic faults have been mapped and studied in detail (Einarsson and Eiriksson 1982; Bjarnason *et al.* 1993; Bergerat and Angelier 2003; Bergerat *et al.* 2003; Angelier *et al.* 2004a; Bellou *et al.* 2005; Clifton and Einarsson 2005; Bergerat *et al.* 2011). Their traces consist of fractures of various length order arranged in steps to the left.

Despite different states of preservation, they all show identical multiscalar arrangement, including the main fault, fault segments, fracture arrays and individual fractures (Figure 2.40). In all cases observed, the main fault is not visible at the surface; it is located at a depth below the first lava flow(s).

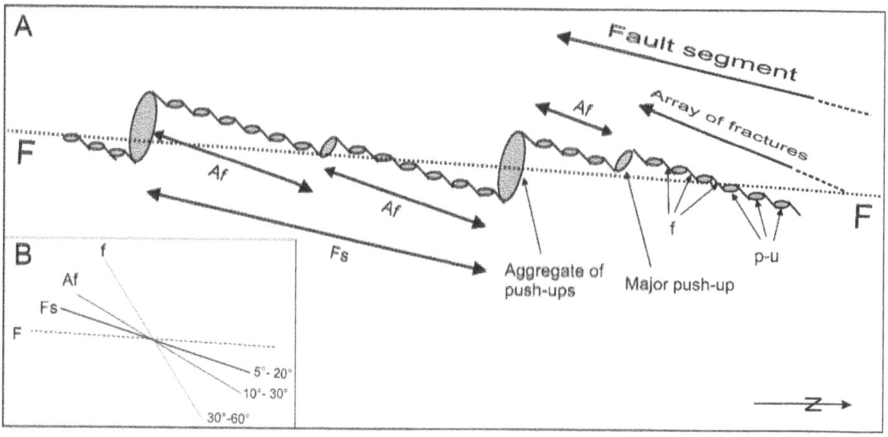

Figure 2.40. *Schematic representation (without scale) of a seismic fault trace in the SISZ (modified from Bergerat* et al. *2011)*

COMMENT ON FIGURE 2.40.– *(A) Illustration of push-up structures (p-u), fractures and multiscalar segmentation with fault segments (second order), arrays of fractures (third order) and individual fractures (fourth order). The F fracture (first order) at depth is indicated by a dotted line. (B) Mean angular relationships between fault (F), fault segment (Fs), array of fractures (Af) and individual fracture (f).*

The lengths of the segments vary according to the faults considered; however, with a few exceptions, they are about 1–2 km. The lengths of the arrays of fracture are mainly around 50–200 m. The width of the steps between fracture groups is proportional to their length. The segments have average angles of 5°–20° with respect to the general direction of the fault and the fracture arrays have average angles of 10°–30° with respect to the segments.

Table 2.1 shows the lengths, directions and angular relationships of these structures for different seismic faults in the SISZ. Various sizes of push-ups accommodate movement in the relay zones, the largest ones being located between fault segments. Sinistral branches, of lesser extension, are sometimes observable, grafted on the main segments or at their extremities. These grafts are generally located where the dextral segments present the maximum displacements, suggesting that an accumulation of deformation is occurring at certain nodes along the faults, promoting the development of conjugate fracture systems. Such sinistral branches are also observed at the end of the dextral segments, which suggest possible small block rotations.

Seismic fault	DS (°)	NS	SL (m)	SS (°)	NA	SM	AL (m)	AS (°)	Nif	o (%)	s (%)	vo (%)
Leirubakki 1294 or 1732?	010°	5	990–2 600	015-035°	11	D	100–200	020–042°	95	17	16	67
					1	S	90	067°				
Selsund 1912	005°	5	137–1 850	010-029°	11	D	41–768	014–032°	290	62	22	16
					2	S	100–206	055–070°				
Minnivellir 1630	005°	4	825–1,700	007-020°	5	D	276–850	002–020°	589	79	21	0
					1	S	562	045°				
Réttarnes 1294 or 1732?	?	1	632	010°	4	D	110–227	022–036°	38	18	24	58
					2	S	47–83	060–070°				
Tjörvafit 1294 or 1732?	?	1	902	025°	5	D	67–412	028–037°	33	9	61	30
					3	S	60–180	057–068°				

Table 2.1. *Table of trace measurements of some historical seismic faults of the SISZ (based on (Bergerat* et al. *2011))*

COMMENT ON TABLE 2.1. – *FS: general fault strike; NS: number of segments; SL: segment lengths; DS: dextral segments strikes; NA: number of fracture arrays; SM: sense of movement (D: dextral, S: sinistral); AL: fracture array lengths; AS: fracture array strikes; Nif: number of individual fractures; o, s and vo: percentage of fractures open, simple and with vertical offset, respectively.*

Figure 2.41. *The Réttarnes Fault (SISZ). (A) Mapping of a segment of the fault (modified from Bergerat* et al. *2011). In gray: push-ups. (B) View toward the north, arrangement of push-ups and fractures illustrating the right-lateral movement of the fault (Françoise Bergerat©)*

Figures 2.41 and 2.42 illustrate two of these faults located in the eastern part of the SISZ: the Réttarnes Fault and the Leirubakki Fault. Their ages are not known with certainty, but Thorodssen (1899) reported a strong earthquake in 1294 in this sector. According to Einarsson (2010), the Leirubakki Fault is the most likely seismic fault to have caused it, although the Réttarnes and Tjörvafit Faults cannot be excluded (Figure 2.39). It is also known that a major earthquake caused significant damage in 1732 in the same sector and the area of damage mapped by Björnsson in 1978 (report of the Civil Defence Working Group, cited by Einarsson 2010) also makes one of these three faults the likely source of this event.

The traces of historical (or prehistoric) seismic faults in the SISZ are more or less preserved on the ground, partly depending on their age, but mainly on the nature of the substratum. While *pahoehoe*-type lavas[3] are favorable for the successful conservation of seismic fault traces even several hundred years after their occurrence

3 *Pahoehoe* and *aa* are Hawaiian terms (used in Iceland) for lava flows with a soft, wavy or corrugated surface (*pahoehoe*) or with a rough, jagged surface (*aa)* (see Chapter 1 of Volume 2).

(Figures 2.41 and 2.42), such traces are, in contrast, very difficult to follow in *aa* lavas. But the nature of the bedrock is not the only element to be taken into account; the thickness of the surface layers (soils) and the presence or absence of vegetation cover are also decisive factors for the types of fractures and their lengths, or for the shape of the push-ups (Figure 2.43), lava without vegetation cover offering, on the whole, better observation conditions.

Figure 2.42. *The Leirubakki Fault (SISZ). (A) Aerial view toward the south (Águst Guðmundsson©, London). (B) Same photograph with indication of the main en échelon fault segments (solid lines) and the main push-ups and push-up aggregates (schematized). The general dextral movement of the fault is indicated*

Push-ups are common along seismic fault traces in the SISZ (Einarsson and Eiriksson 1982; Bjarnason *et al.* 1993; Bergerat *et al.* 2011). They result from a compression process between blocks located on both sides of the fractures they connect. Their shortening thus depends on the value of the horizontal offset of the deep strike-slip fault. They only affect the lava layers close to the surface, where the *en échelon* fractures develop. Their analysis thus allows for reconstructing both the displacement along the faults and the depth of the detachment zone above which the fault is no longer expressed by a single plane.

Figure 2.43. *Different shapes of push-ups according to the substratum (modified from Bergerat* et al. *2011; Françoise Bergerat©)*

COMMENT ON FIGURE 2.43.– *(A) Push-up affecting basaltic lavas with little or no surface cover (on the Leirubakki Fault). (B) Push-up affecting lava layers with relatively thick tuff and/or soil cover (on the Selsund Fault). (C) Push-ups formed in a very thick cover and showing a "crater" shaped depression at their tops (on the Minnivellir Fault). (D) Small push-up affecting the soil of a meadow (on the Árnes Fault).*

The shortening ΔL, calculated in a direction perpendicular to the axis of the push-up, is given by the difference between the final length L between two points and the initial length L_0, which is the envelope of the push-up (Figure 2.44A):

$$\Delta L = L_0 - L \quad [2.3]$$

The depth at which the *en échelon* arrangement is replaced by a single fault can be evaluated according to the volume of the push-up:

$$H = S/\Delta L \quad [2.4]$$

In fact, since the different segments (FS) are arranged *en échelon* and therefore slightly oblique with respect to the general direction of the fault (F), one must slightly correct this value according to the angle (Φ) between FS and F (Figure 2.44B). This makes it possible to determine the actual displacement (D) for each push-up along the fault line and, finally, all the displacements along the fault line:

$$D = \Delta L \cos \Phi \qquad [2.5]$$

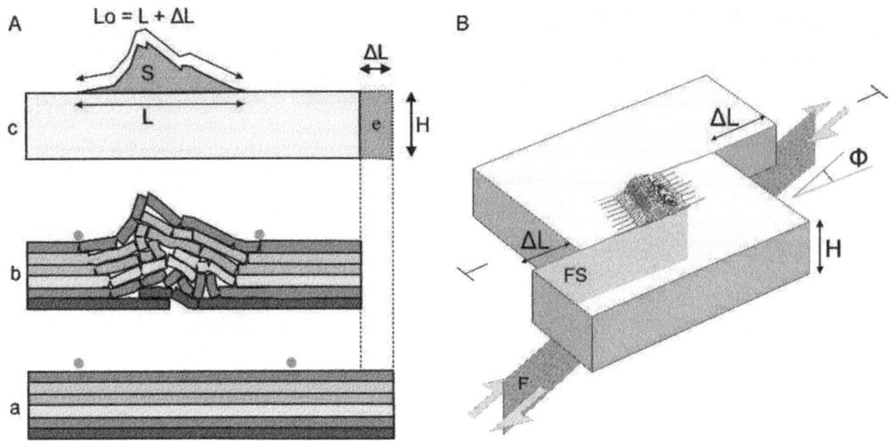

Figure 2.44. *Geometrical characteristics of push-ups and associated fractures (modified from Bergerat* et al. *2003)*

COMMENT ON FIGURE 2.44.– *(A) Determination of shortening and depth of detachment through a push-up structure. Sections perpendicular to the push-up axis: before the earthquake (a) and after the disorganization of the lava layers due to the earthquake (b); in (c), schematic section of stage b (along the trace indicated on the block diagram in (B) showing the different elements measured and/or calculated. (B) Relationships between the geometry of fractures and push-ups at the surface and at depth. L: horizontal distance between two points located on either side of the push-up, at its base (gray points in a and b); L0: length of the push-up between these two points; ΔL: shortening perpendicular to the push-up axis; S: volume of the push-up per unit length; e: thickness of the affected lava layers; Φ: angle between the direction perpendicular to the push-up axis and the strike-slip fault.*

Several authors (Wells and Coppersmith 1994; Stirling *et al.* 2002; Kim and Sanderson 2005) have proposed empirical relationships between certain parameters of seismic faults, in particular between magnitude and displacement (maximum or average) along the fault. Using the linear regression laws thus provided, the magnitude of an earthquake can be calculated from push-up measurements.

In the case of the faults illustrated above (Figures 2.41 and 2.42), the application of the linear regression laws established by Wells and Coppersmith (1994) for the strike-slip faults indicates a magnitude of 7.05–7.06 for the earthquake at the origin of the Réttarnes Fault and 7.09–7.15 for the one related to the Leirubakki Fault (Table 2.2; Bergerat *et al.* 2011).

The same types of calculations made from push-up measurements on the Selsund Fault, whose M_S magnitude was estimated at 7.1 by Karnik (1969) from instrumental records, gave a magnitude of 7.1 ± 0.15 (Angelier *et al.* 2004a), thus confirming the validity of the determinations made on the basis of these field studies.

Seismic fault	ML (km)	MD (m)	AD (m)	M	TL (km)
Tjörvafit	0.9	1.96	1.27	7.03–7.13	47–57
Réttarnes	0.6	2.07	1.08	7.05–7.06	49–50
Minnivellir W + N + E	5.2	4.54	2.32	7.32–7.36	85–92
Minnivellir W + N	5.2	2.7	1.32	7.14	59
Minnivellir E	0.9	1.84	0.97	7.01–7.02	45–46
Leirubakki	8	2.78	1.14	7.09–7.15	53–60
Selsund	9	2.4	1.7	7.1–7.24	54–72

Table 2.2. *Determination of the total magnitude and rupture length of some historical seismic faults in the SISZ based on displacement calculated from push-up analysis (based on Bergerat* et al. *2011)*

COMMENT ON TABLE 2.2.– *Two calculations are given for the Minnivellir Fault, depending on whether its two branches are considered to correspond to one (italicized in the table) or two distinct earthquakes (branches W and N, branch E). LT: length of the visible trace; MD: maximum displacement; AD: average displacement; M: range of magnitude estimated from MD and AD; TL: total length of the rupture calculated from the magnitudes.*

The magnitudes estimated for historical seismic faults whose traces are visible in the SISZ thus correspond to major seismic events. Their use to estimate total dimensions provides fault lengths of 45–70 km, demonstrating that significant portions of fault traces are no longer – or hardly – observable. These calculations also reveal that events significantly larger than those of June 17 and 21, 2000 (Ms = 6.6; Stefánsson *et al.* 2000) have occurred in the relatively recent past, which should be taken into account in the seismic risk assessment.

2.3.2.2.3. Models for the SISZ

Despite the general E-W direction of SISZ, the majority of earthquakes occur on N- S faults (Stefansssson *et al.* 1993; Rögnvaldsson and Slunga 1994). This is also the direction of traces of historical or older earthquakes observable in Holocene lavas (Jóhannesson *et al.* 1990; Einarsson *et al.* 2005), although in both cases the presence of minor sinistral fault segments is observed. This arrangement indicates that the transform movement is not accommodated by one or more major sinistral strike-slip faults, but essentially by dextral strike-slip motions occurring on submeridian faults.

Guðmundsson and Brynjolfsson (1993) explained this particular geometry by the stress field existing in the overlapping rift-zone segment.

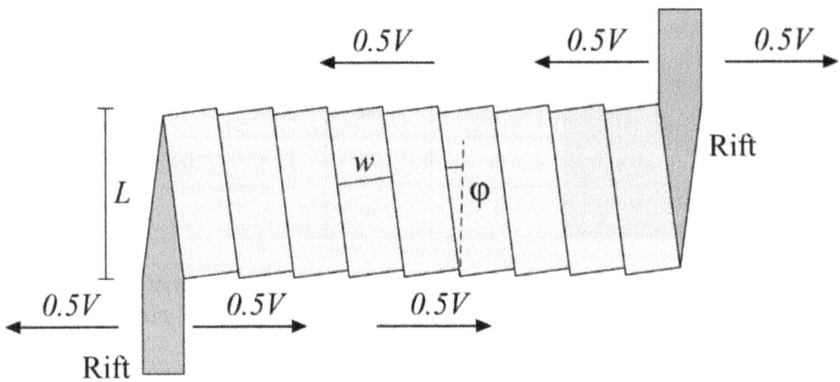

Figure 2.45. *Schematic representation of bookshelf faulting in the SISZ (modified from Sigmundsson* et al. *1995; Sigmundsson 2006)*

The fault arrangement in the SISZ has been interpreted by some authors (Einarsson *et al.* 1981; Einarsson and Eiriksson 1982; Hackman *et al.* 1990; Einarsson 1991; Sigmundsson *et al.* 1995) as an example of bookshelf faulting and by others (Bergerat *et al.* 1999; Angelier and Bergerat 2002) as a Riedel-type shear zone (Riedel 1929).

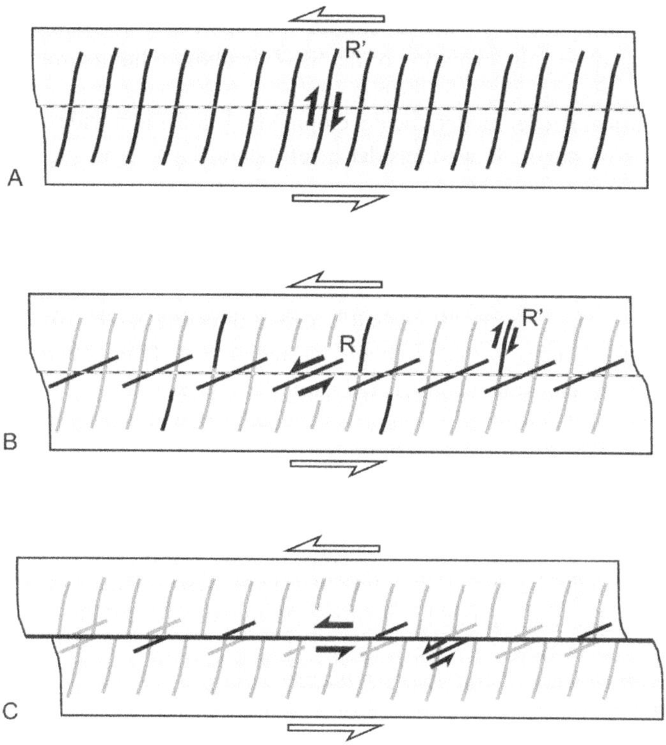

Figure 2.46. *Theoretical development of a major E-W sinistral strike-slip zone, following a Riedel-type model (modified from Angelier* et al. *2004a). The active faults are shown in black at each step*

In the bookshelf faulting model (Figure 2.45), the deformation is accommodated by a series of faults parallel to each other and perpendicular to the shear direction of the SISZ (i.e. N-S), like books on a shelf pushed sideways. The slip on each fault depends on the divergence velocity of the plates (V), the width of the zone (L) and the distance between the faults (w); the blocks between the faults will rotate at an angle $\varphi = \tan^{-1}$ (V/L). In the SISZ, the rotation rate would thus be about 0.25 µrad/year and the average slip rate on each fault ($s = w \tan \varphi$) would be 0.5 mm/year (Sigmundsson *et al.* 1995; Sigmundsson 2006).

This model has been questioned (Guðmundsson 1995b; Angelier and Bergerat 2002) because to maintain such a structure throughout the entire history of the SISZ would imply significant rotations, and deformations, which is not in agreement with the observations. Such a model cannot be excluded if assuming frequent reorganizations during the evolution of the SISZ, leading to limited rotations.

In the classical Riedel model, the development of faults in a wide strike-slip zone occurs in several stages. If we consider a left-lateral E-W zone such as the SISZ, initially, only strike-slip faults in the opposite direction to the main shear and very oblique with respect to the direction of the latter develop. These are the "Riedels R'", here trending N 10°–20° E and dextral (Figure 2.46(A)). Then, as the general shear continues, faults in the same direction as the main movement and slightly oblique to its direction appear. These are the "Riedels R", here of direction N 60°–70° E and sinistral (Figure 2.46(B)). Finally, at the last stage, the major E-W fault is expressed (Figure 2.46(C)). The mechanisms reconstructed from the traces of historical seismic faults and the focal mechanisms of earthquakes indicate that the recent to present-day functioning of the SISZ corresponds to an intermediate stage between (A) and (B) with a majority of dextral faults and the development of the few sinistral faults.

The origin of the fault pattern is a question that remains debated to this day. However, it should be noted that the two hypotheses (Riedel-type model and bookshelf model) are not mutually exclusive (Bergerat *et al.* 2011).

2.3.2.3. *Comparison of the two transform zones in terms of geometry and stress state*

The two transform zones, TFZ and SISZ, do not reach the same stage of development. The geometrical configuration of the faults that constitute them, as well as the stress fields reconstructed from the movements of these faults, reflect this different degree of maturity. Even within each of these two transform zones, the fault zones may show different degrees of development.

It is thus possible to observe various fault systems, from zones of diffuse deformation without major surface faults to zones where these faults are well individualized and of the same direction as that of the plate movement (Bergerat and Angelier 1999, 2008).

The deformation in the most recent transform zone, the SISZ, can be described as diffuse. There are no major E-W oriented sinistral faults and the fault geometry corresponds to a classical Riedel-type model, with dextral N 0°–20° E and sinistral N 60°–70° E strike-slip faults. The dominant stress field in this zone is characterized by a NW-SE extension direction, in agreement with the sinistral transform motion (Figure 2.47(A)). The SISZ can thus be considered as an immature transform zone.

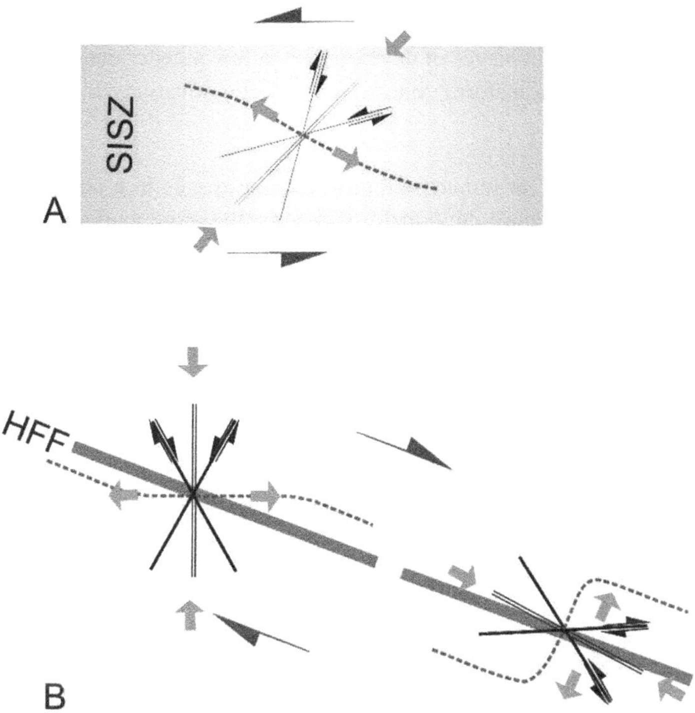

Figure 2.47. *Schematic representation of faults and corresponding stress states in the SISZ (A) and along the HFF (B) (modified from Bergerat and Angelier 2008)*

COMMENT ON FIGURE 2.47.– *Strike-slip faults: solid lines with indication of shear sense; normal faults: double lines; directions of compression and extension: gray arrows. The two groups of faults and stress states along the HFF do not correspond to different locations, but illustrate the two major stress regimes (details in Figure 2.34).*

In the TFZ, the three major lineaments show different stages of development. The Dalvík lineament is also a zone of diffuse deformation including NNE-SSW seismic faults (section 2.3.2.1). This is as well the case for the GL, which is not observable on land but whose seismicity leads to NNE-SSW faults (Rögnvaldsson *et al.* 1998). Only the HFF is clearly marked by a major ESE-WNW dextral strike-slip fault that can be followed on land for more than 70 km. In contrast, the stress field is complex, with two dominant extension directions, E-W and NE-SW. These two extensions are compatible with the dextral transform motion and reveal different coupling behaviors (Figure 2.47(B)). The TFZ thus includes faults at different stages of maturity, this arrangement resulting from both the northward migration of the NVZ and the southward migration of the KR.

These different expressions of deformation related to the transformation process, and observable at distinct stages of development, allow a better understanding of the evolution of Icelandic transform zones (Bergerat and Angelier 2008; Karson *et al.* 2018).

Generally speaking, an immature transform zone results in a complex structural pattern and the more mature the transform zone is, the more marked the expression of the surface fault becomes.

On the contrary, the stress field that prevails in the early stages of the transform process seems quite simple, whereas in a mature zone, it can be complex, with frequent stress permutations. It thus leads to coupling and decoupling processes, alternately along major faults.

2.4. Geothermal energy and hydrothermalism

2.4.1. *Geothermal systems*

Iceland's location on both an ocean ridge and a hot spot results in high heat flow and geothermal activity with regional heat flow values ranging from 80 to 150 mW.m^{-2}. The most significant heat flow measurements are located near the active rift and decrease with increasing distance from it (Flóvenz and Sæmundsson 1993). They are much higher than those measured on the continents (about 40–70 mW.m^{-2}) (Sclater *et al.* 1980).

The classification of geothermal systems into high temperature (HT) and low temperature (LT), commonly used in Iceland, dates back to about 60 years (Bödvarsson 1961; Sæmundsson *et al.* 2009). It is arbitrarily based on the temperature at a depth of 1 km, with HT fields being those where the temperature reaches 200°C and LT fields those where the temperature is below 150°C. The HT and LT geothermal regions are thus, respectively, located in the volcanic zones (rift) and outside them (Figure 2.48).

Classically, geothermal areas located around active volcanoes (Leifsson 1992; Flovenz and Sæmundsson 1993; Arnórsson 1995a) are considered HT geothermal fields (Figure 2.49(A)). These geothermal systems include deep aquifers heated by contact with ambient hot rocks and controlled by fractures. Icelandic geothermal systems in the intermediate range (150°–200°C) are rare (Sæmundsson *et al.* 2009), but the boundary is not always clear and some areas, such as Hveragerði in the south of the island (Figures 2.49(B) and (C)), are sometimes considered transitional between these two types (Arnórsson 1995b).

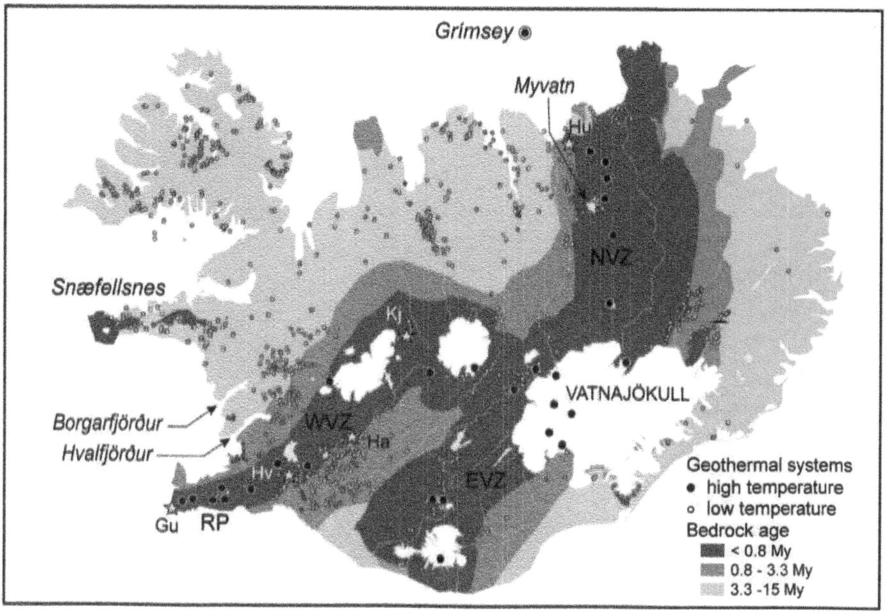

Figure 2.48. *Distribution of low- and high-temperature geothermal systems in Iceland (modified from Axelsson* et al. *2010). The location of the geysers discussed in section 2.4.2 is indicated by yellow stars on the map. Gu: Gunnuhver; Ha: Haukadalur; Hv: Hvergarði; Hu: Húsavík; Kj: Kjölur*

The LT geothermal fields belong to the Tertiary and Lower Pleistocene age regions (Figure 2.48), the main ones being located in southwest Iceland. The temperature of their reservoirs varies from ambient temperature to about 150°C (Arnórsson 1995b). They are often revealed by the presence of hot springs (flow rate between almost zero and 180 L/s). However, in the last decades many of them have been detected, even though they were not expressed at the surface (Flovenz and Sæmundsson 1993), because of shallow drilling campaigns allowing the geothermal gradient to be determined and thus characterizing the shape and apex of the thermal anomaly (Figure 2.50).

The Deildartunga hot spring, shown in Figure 2.51, is a typical example of the use of LT geothermal energy in the west of the island.

Figure 2.49. *Use of high-temperature geothermal energy in southwest Iceland. (A) Geothermal site near the Hengill volcano, WVZ. (B) Anthropogenic geyser on a steam line. (C) Steam collecting area in Hveragerði (Brigitte Van Vliet-Lanoë©)*

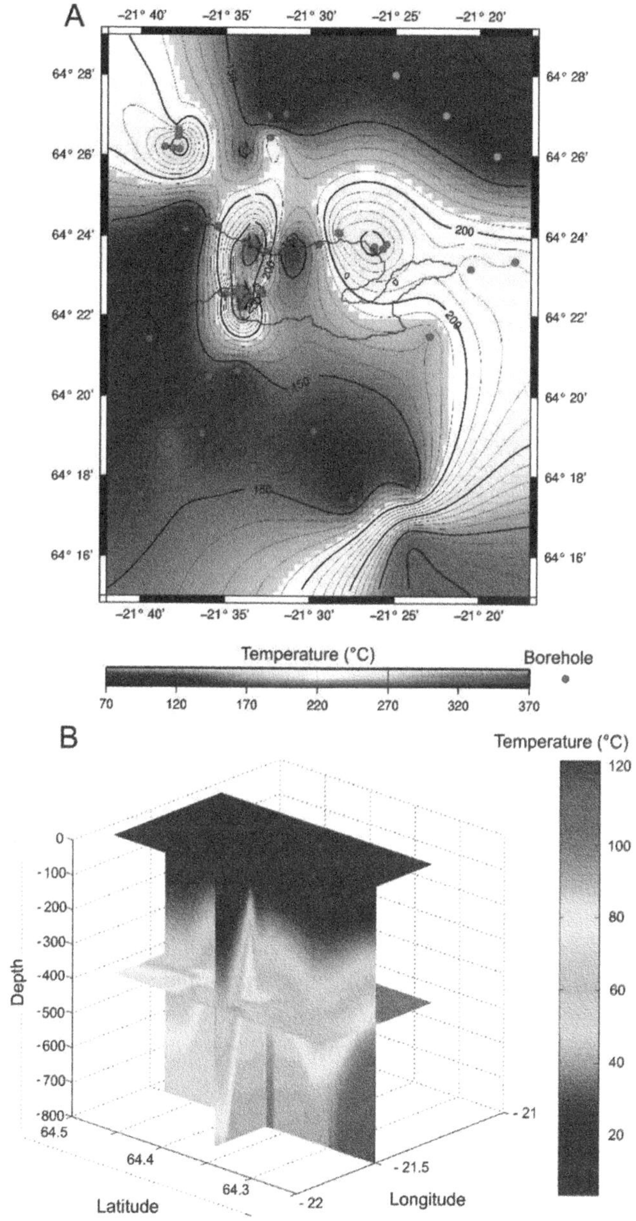

Figure 2.50. *The thermal anomaly of the Hvalfjörður. (A) Map of the geothermal gradient established from 57 drill holes. (B) Temperature model (according to Angelier* et al. *2004b)*

Figure 2.51. *The Deildartunga (Deldartunguhver) hot spring in western Iceland (Borgarfjörður). Steam and boiling water at the outcrop (A) and greenhouses with tomato cultivation (B, C) next to the geothermal station (Françoise Bergerat©)*

COMMENT ON FIGURE 2.51.– *Deldartunguhver is the most significant hot spring in Iceland (and even in Europe) in terms of volume. Its average flow rate is 180 L/s of water at 100°C. It has been in use since 1925 and is currently managed by the Akranes-Borgarfjörður district heating company. The station is located 19 m above sea level and extraction is by pumping. The pipeline system connecting it to Akranes, Borgarnes and their surroundings was built between 1979 and 1981 and has a total length of 74 km. It is made of asbestos-cement and in places of steel. The hot water takes about a day to reach Akranes and its temperature drops by about 20 degrees on the way. The average temperature on arrival in Borgarnes (34 km to the SE) is 77°C and in Akranes (64 km to the SE) 73°C.*

The simple conceptual model of a fracture-controlled geothermal field is based on the presence of subvertical fractures allowing the development of convective cells. Meteoric water infiltrates and migrates downward along various paths (cooling joints, faults, dykes, etc.) in the host rock where, at a depth of a few kilometers, it is heated and then rises to the surface (Figure 2.54).

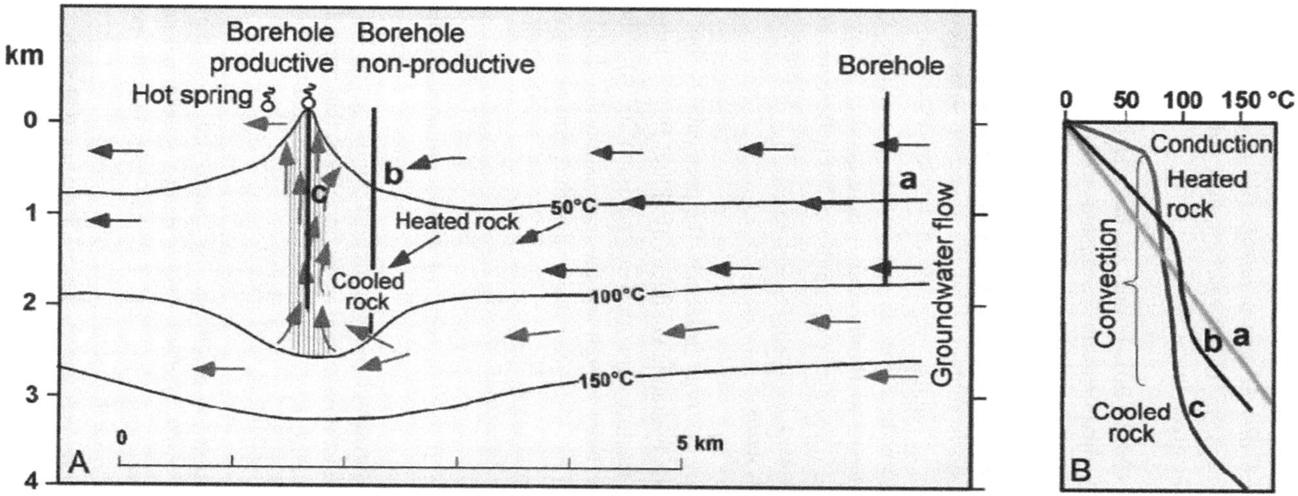

Figure 2.52. *Simplified model of a low-temperature geothermal field (modified from Bergerat* et al. *2013)*

COMMENT ON FIGURE 2.52.– *(A) Characteristics of a fracture-controlled geothermal system. The arrows indicate the migration of cold (blue) and warm (red) water. (B) Thermal state of the system. Geothermal gradient outside (green) and inside (red and blue) the system. The green line is the average regional geothermal gradient.*

For Sæmundsson (2011, 2014), one of the mechanisms likely to generate low-temperature geothermal activity is related to the extent of dyke swarms belonging to the active volcanic systems, far away from the latter (up to more than 50 km). These dykes can propagate at depth far beyond surface expressions such as nearby fissural eruptions, cracks (*gjá*) that extend further and hot spring zones (even further). This hypothesis is supported by the shape of some geothermal gradient profiles showing peaks at regular intervals of 4–5 km (Sæmundsson 2014). Thus, dyke swarms issued from the Krýsuvík and Svartsengi volcanic systems of the RP could extend northeastward to Hvalfjörður, where geothermal systems are present while there is no active volcanism (Sæmundsson 2011; Bergerat *et al.* 2013). These dykes do not then act as a heat source, but generate a vertical permeability and, therefore, convection systems.

The fracture network not only influences the permeability, but also introduces a certain anisotropy in the hydraulic conductivity. Fluid flow is largely controlled by the permeability of its fracture network (Guðmundsson 2000a). When there are many interconnections, permeability is high and the percolation threshold is easily reached (Philipp *et al.* 2007). It has also been shown that the permeability of faults increases if they are active or, on the contrary, is low if they are inactive (Guðmundsson *et al.* 2002).

Nevertheless, even if the analysis of brittle structures helps to understand geothermal processes on a regional scale, such analysis on a local scale is ultimately of minor interest in the search for geothermal fields, given the polyphased and complex context of faults, mineralized veins and dykes that prevails in Iceland. The most efficient tool remains the geothermal survey from shallow boreholes (Flovenz and Sæmundsson 1993; Bergerat *et al.* 2013).

While the production of electricity for industrial use (particularly for the aluminum industry) is mostly hydroelectric in nature, geothermal energy plays a key role in the Icelandic economy. HT and LT resources provide nearly 70% of the domestic energy supply for the island's 320,000 inhabitants. Currently, most of the country's district heating uses energy from LT geothermal systems located outside the volcanic zone. In most cases, hot water is collected by pumping, but there are examples of artesian rises that are still in operation.

There are currently 22 public or municipal heating companies in Iceland, managing the heating networks (*hitaveitas*) of 62 districts. Many of these *hitaveitas* have been operational for several decades, the oldest being over 80 years old (Axelsson *et al.* 2010). Reykjavík's geothermal district heating service is by far the largest. It began operating in 1930 and serves more than 180,000 people in Reykjavík and five surrounding municipalities, about 58% of Iceland's total population (Axelsson *et al.* 2010). The hot water for this network comes from five geothermal sectors, three of LT and two of HT (Gunnlaugsson and Ívarsson 2010). In rural areas, there are also small private networks serving 10–20 farms (currently about 4,000 inhabitants). Geothermal energy is also used for horticulture (heating of greenhouses; see Figures 2.53(B) and (C)), fish farming, urban street snow removal systems, swimming pools – of Iceland's 160 swimming pools, 130 are heated by geothermal energy – or as a part of the electricity production (Ragnarsson 2008).

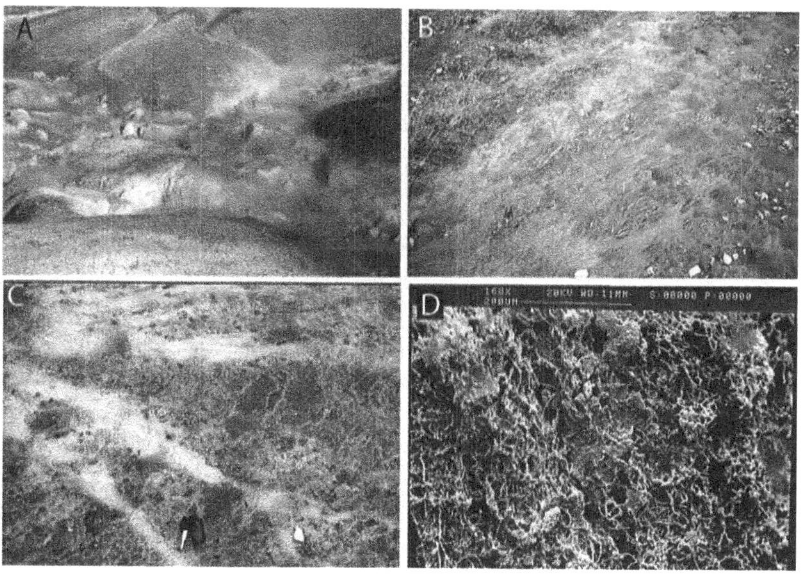

Figure 2.53. *Products of hydrothermal activity. (A) Seltún hot springs area (Kleifarvatn). (B) Filamentous algae in a very hot flow zone (cyanobacteria) (Geysir). (C) Bacterial opale precipitation in microgours (Geysir). (D) Scanning electron microscope photograph showing a film of encrusting cyanobacteria (opal) (Geysir) (Brigitte Van Vliet-Lanoë©)*

Hydrothermal sources, from a geochemical point of view, can be acidic or alkaline (Figure 2.53). The presence of corrosive hot water, rich in CO_2 and often other products (H_2S, H_2SO_4, chlorides), induces a significant alteration of the volcanic substrate or the metallic pipes used to conduct the water for anthropogenic applications. The volcanic glass is altered first to red ferrous clays (chlorites and nontronites), then to gray clays (reduced and deferrified) with halloysite mixed with kaolinite (acidic medium), or kaolinite alone (alkaline medium). They release in waters silica, calcium and iron hydroxides (Browne 2003; Tómasson and Kristmannsdóttir 1972). These transformations make the surroundings of springs very muddy or superficially encrusted with gypsum or sulfur deposits.

Deposits of opal (geyserite) (Tobler *et al.* 2008; Campbell *et al.* 2015) and biogenically precipitated carbonates at temperatures >40°C can form basins in flow areas or on the edges of geyser vents. The "blue", alkaline waters contain particulate silica products (opal, diatomaceous earth) when their temperature is <40°C. At higher temperatures, the silica remains solubilized. Hot springs and geysers can shelter archaeobacteria adapted to very hot water and reducing conditions. The red, yellow, blue and green colors of pools, the white filaments or the dark-green sticky *mattes* surrounding the hot basins are actually due to the occurrence of thermophilous organisms such as bacilli or types of cyanobacteria (*Mastigocladus laminosus*, *Chloroflexus*) living in the sulfurous and superheated water (Stetter 1996; Takacs *et al.* 2001).

2.4.2. *Geysers and hydrothermalism*

A geyser is a particular type of hydrothermal water source that gushes out intermittently but relatively regularly by throwing up water and steam at HT and high pressure, and sometimes small stones or mud, swept over the sides of the active pipe (Figure 2.54). There are several such sites in Iceland, most of them in the WVZ. The main site is located in the Haukadalur valley (Great Geysir and Strokkur), but some can also be found in the town of Hvergarði (Grýla), in Hveravellir, near the Kjölur trail in the north-east of Langjökull, at another site called Hveravellir (Ystihver, Uxahver and Syðstihver) south of Húsavík. A peculiar one was in activity in the past at Gunnuhver, blowing seawater, at the southern tip of the RP (see location in Figure 2.48).

Meteoric or marine water seeps into the geyser reservoir near a magma reservoir. It will become charged with salts and dissolved gases in contact with this reservoir, accumulate, then will rise upwards along a fracture forming a constrained conduit. The hydrostatic pressure will increase the boiling temperature. The deeper the reservoir is, the higher the pressure and the boiling temperature. After a certain time, the water reaches a supercritical state; vapor bubbles form at depth and then rise to the surface through the conduit, pushing the water upwards. As it rises, the bubble reduces the hydrostatic overpressure and allows the water to boil. The water is then expelled to the surface through the existing conduit. The eruption ends with the partial emptying of the water column, when the bubble reaches the surface and explodes, releasing the accumulated steam: this is what forms the geyser (Adelstein *et al.* 2014).

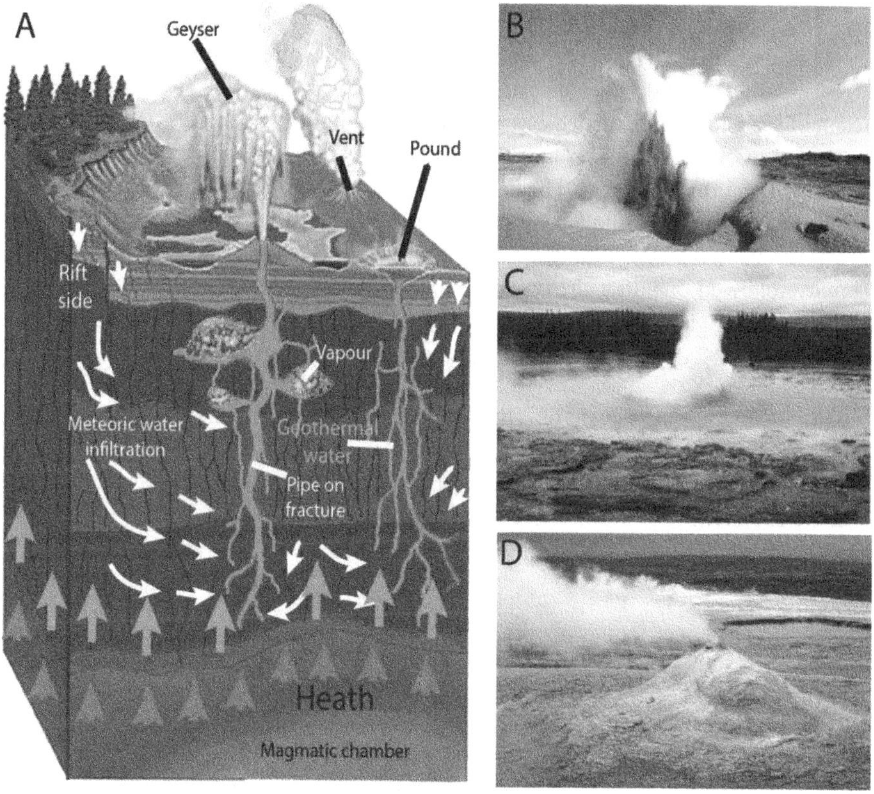

Figure 2.54. *Geysers and hydrothermal sources*

COMMENT ON FIGURE 2.54.– *(A) Schematic diagram of a geyser (modified from* Encyclopedia Britannica*). (B) Geyser with conduit sweeping and sludge (sea water) emission, Gunnuhver, southern tip of the Reykjanes Peninsula. (C) Bubble exploding at the beginning of the geyser's gushing, Strokkur, Geysir site. (D) Geyserite vent on a currently non-functional geyser, Kjölur, northeast of Langjökull (photos B, C, D: Brigitte Van Vliet-Lanoë©).*

A new eruptive cycle (or eruptive sequence) resumes when enough meteoric water has re-entered the conduit for the water to rise again, expand and begin vaporizing once more. The duration of each eruption and the time between eruptions vary from one geyser to another, and their regularity is also variable. The intensity of the ejection logically increases with a decrease in atmospheric pressure.

This explains why geyser geothermal zones are often aligned along fractures (Georgsson *et al.* 2005; Neale *et al.* 2016; Walter *et al.* 2018). These can change as a result of seismic activity, which can alter the geometry of the conduits and the pressure in the reservoirs (Stefánsson *et al.* 2000; Björnsson *et al.* 2001). The Great Geysir was first mentioned in 1294. The earthquake of 1630 brought its activity to a halt for 40 years until a new earthquake in 1670. Thereafter, it hardly gushed out anymore until it was reactivated by the earthquake of 1896. Periods of reduced activity and of intense activity followed one after the other. Following the earthquakes of June 2000, Geysir began to erupt again after having been more or less inactive since 1935. It gushed out 122 m high for 2 days, making it one of the highest geysers in the world.

Some hot springs gush out like geysers from the boreholes and watertight conduits established by man at geothermal exploitation sites. These geysers are artificial and associated with geothermal wells (254 m, 90°C) such as at Hvergarði (Figure 2.49(B)).

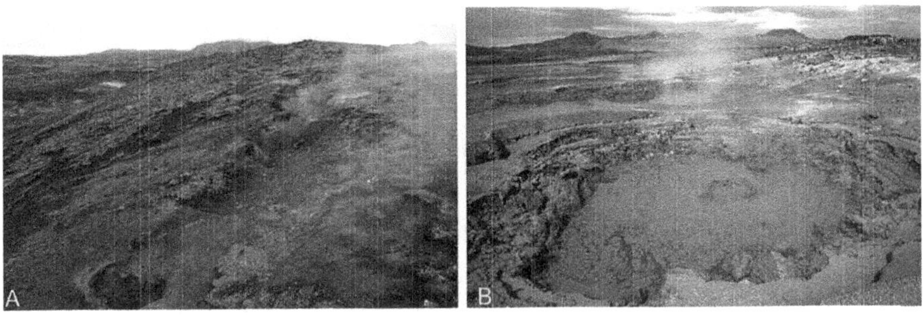

Figure 2.55. *Hydrothermal activity at Námafjall, in the Myvatn sector (NVZ). (A) Vents and fumaroles along a fracture (Brigitte Van Vliet-Lanoë©). (B) Solfatares and mud pots (Françoise Bergerat©)*

In Iceland, the most spectacular hydrothermal phenomena are springs and geysers. However, many areas of fissures and vents, generally referred to as fumaroles, also emit water vapor, hydrogen sulfide, carbon dioxide and other gases (Figure 2.55(A)). When hydrogen sulfide (H_2S) partially oxidizes with atmospheric O_2, it evolves into sulfur vapors, forming beautiful crystalline crusts near the vents.

Solfatares emit water in vapor form and various gases: CO_2, CH_4, NH_4 and H_2S. In addition to sulfur compounds and sulfates, clays are also present. These various deposits give magnificent yellow, ochre and red hues as well as a characteristic foul smell of rotten eggs. The solfatares are characterized by the simultaneous presence of a liquid phase, a gaseous phase, alteration products (clays) and sulfur. Boiling, steaming and foul-smelling mud potholes can form in solfatares where water is abundant (Figure 2.55(B)). The clay mixes with the water to form a more or less viscous mud, with bubbles of carbon dioxide rising to the surface and bursting. The mud is thrown up on the edges of the ponds, forming small craters. The most remarkable solfatare fields are in the NVZ (Námafjall, Þeistareykir, Krafla) and the RP (Seltún).

2.5. References

Acocella, V., Guðmundsson, Á., Funiciello, R. (2000). Interaction and linkage of extension fractures and normal faults: Examples from the rift zone of Iceland. *J. Struct. Geol.*, 22, 1233–1246.

Acocella, V., Korme, T., Salvini, F. (2003). Formation of normal faults along the axial zone of the Ethiopian Rift. *J. Struct. Geol.*, 25, 503–513.

Adelstein, E., Tran, A., Munoz, C., Steinberg, A., Manga, M. (2014). Geyser preplay and eruption in a laboratory model with a bubble trap. *J. Volcanol. Geotherm. Res.*, 285, 129–135 [Online]. Available at: http://doi.org/10.1016/j.jvolgeores.2014.08.005.

Ágústsson, K., Stefánsson, R., Linde, A.T., Einarsson, P., Sacks, I.S., Guðmundsson, G.B., Þorbjarnardóttir, B. (2000). Successful prediction and warning of the 2000 eruption of Hekla based on seismicity and strain changes. *AGU Fall Meeting*, San Fransisco, 1337.

Angelier, J. and Bergerat, F. (2002). Behaviour of a rupture of the 21 June 2000 earthquake in South Iceland as revealed in an asphalted car park. *J. Struct. Geol.*, 24, 1925–1936.

Angelier, J., Bergerat, F., Dauteuil, O., Villemin, T. (1997). Effective tension-shear relationships in extensional fissure swarms, axial rift of northeastern Iceland: Morphological evidences. *J. Struct. Geol.*, 19, 673–685.

Angelier, J., Bergerat, F., Homberg, C. (2000). Variable coupling across weak oceanic transform fault: Flateyjarskagi, Iceland. *Terra Nova*, 12, 97–101.

Angelier, J., Bergerat, F., Bellou, M., Homberg, C. (2004a). Co-seismic strike-slip displacement determined from push-up structures: The Selsund Fault case, South Iceland. *J. Struct. Geol.*, 26, 709–724.

Angelier, J., Bergerat, F., Sæmundsson, K., Fourel, L., Jónsson, S.S. (2004b). Geothermal research, fracture system analysis and paleostress reconstruction: A case study at Hvalfjörður, Iceland. *20e Réunion des Sciences de la Terre*. Société Géologique de France-Geologische Vereinigung, Strasbourg.

Angelier, J., Bergerat, F., Stefánsson, R., Bellou, M. (2008). Seismotectonics of a newly formed transform zone near a hot spot: Earthquake mechanisms and regional stress in the South Iceland Seismic Zone. *Tectonophysics*, 447, 95–116.

Antonioli, A., Belardinelli, M.E., Bizzarri, A., Vogfjörð, K.S. (2006). Evidence of instantaneous dynamic triggering during the seismic sequence of year 2000 in south Iceland. *J. Geophys. Res.*, 111, B03302 [Online]. Available at: http://doi.org/10.1029/2005JB003935.

Arnórsson, S. (1995a). Geothermal systems in Iceland: Structure and conceptual models I. High temperature areas. *Geothermics*, 24(5–6), 561–602.

Arnórsson, S. (1995b). Geothermal systems in Iceland: Structure and conceptual models II. Low temperature areas. *Geothermics*, 24(5–6), 603–629.

Arnórsson, S., Axelsson, G., Sæmundsson, K. (2008). Geothermal systems in Iceland. *Jökull*, 58, 269–302.

Aronson, J.L. and Sæmundsson, K. (1975). Relatively old basalts from structurally high areas in central Iceland. *Earth Planet. Sci. Lett.*, 28, 83–97.

Árnadóttir, Þ., Hreinsdóttir, S., Guðmundsson, G., Einarsson, P., Heinert, M., Völksen, C. (2001). Crustal deformation measured by GPS in the south Iceland seismic zone due to two large earthquakes in June 2000. *Geophys. Res. Lett.*, 28, 4031–4033.

Árnadóttir, Þ., Geirsson, H., Einarsson, P. (2004). Coseismic stress changes and crustal deformation on the Reykjanes Peninsula due to triggered earthquakes on June 17, 2000. *J. Geophys. Res.*, 109(B09307) [Online]. Available at: http://doi.org/10.1029/2004JB003130.

Árnadóttir, Þ., Jónsson, S., Pollitz, F.F., Jiang, W., Feigl, K.L. (2005). Postseismic deformation following the June 2000 earthquake sequence in the south Iceland seismic zone. *J. Geophys. Res.*, 110(B12308) [Online]. Available at: http://doi.org/10.1029/2005jb003701.

Árnadóttir, Þ., Jiang, W., Geirsson, H., Sturkell, E., Pagli, C., Sigmundsson, F., Einarsson, P., Sigurdsson, Þ. (2006). Plate spreading and rapid uplift observed by GPS in Iceland. Eos Trans. *AGU Fall Meeting*. American Geophysical Union, San Francisco, 87(52).

Árnadóttir, Þ., Lund, B., Jiang, W., Geirsson, H., Björnsson, H., Einarsson, P., Sigurdsson, T. (2009). Glacial rebound and plate spreading: Results from the first countrywide GPS observations in Iceland. *Geophys. J. Int.*, 177(2), 691–716.

Axelsson, G., Gunnlaugsson, E., Jónasson, T., Ólafsson, M. (2010). Low-temperature geothermal utilization in Iceland: Decades of experience. *Geothermics*, 39, 329–338.

Bagdasaryan, G.P., Gerasimovkiy, V.I., Polyakov, A.I., Gukaysan, R.K. (1976). New data on the absolute age of Icelandic volcanic rocks. *Geokhimiya*, 9, 1333–1339.

Bellou, M. (2006). Analyse sismotectonique de la zone sismique sud-islandaise. PhD Thesis, Université Pierre et Marie Curie, Paris–Villefranche-sur-Mer, Paris.

Bellou, M., Bergerat, F., Homberg, C., Angelier, J. (2005). Geometry and segmentation mechanisms of the surface traces associated with the 1912 Selsund Earthquake, Southern Iceland. *Tectonophysics*, 404, 133–149.

Bergerat, F. and Angelier, J. (1998). Fault systems and paleostresses in the Vestfirdir peninsula. Relationships with the Tertiary paleo-rifts of Skagi and Snaefells (Northwest Iceland). *Geodin. Acta*, 11(2–3), 105–118.

Bergerat, F. and Angelier, J. (1999). Géométrie des failles et régimes de contraintes à différents stades de développement des zones transformantes océaniques : exemple de la zone sismique sud-Islandaise et de la zone de fracture de Tjörnes (Islande). *C.R. Acad. Sc.*, 329, 653–659.

Bergerat, F. and Angelier, J. (2000). The South Iceland seismic zone: Tectonic and sismotectonic analyses revealing the evolution from rifting to transform motion. *J. Geodin.*, 29(3–5), 211–231.

Bergerat, F. and Angelier, J. (2001). Mécanismes des failles des séismes des 17 et 21 juin 2000 dans la zone dsismique sud-Islandaise d'après les traces de surface des failles d'Árnes et de l'Hestfjall. *C.R. Acad. Sc.*, II(333), 35–44.

Bergerat, F. and Angelier, J. (2003). Mechanisms of the June 2000 earthquakes, southern Iceland: Inferences from surface traces of the Arnes and Hestfjall faults. *J. Struct. Geol.*, 25, 1507–1523.

Bergerat, F. and Angelier, J. (2008). Immature and mature transform zones near a hot spot: The South Iceland seismic zone and the Tjörnes fracture zone (Iceland). *Tectonophysics*, 447, 142–154.

Bergerat, F., Angelier, J., Villemin, T. (1992). Déformations cassantes dans la partie émergée d'une zone transformante océanique : la zone de fractures de Tjörnes (Islande). *C.R. Acad. Sci.*, II(315), 1011–1018.

Bergerat, F., Guðmundsson, Á., Angelier, J., Rögnvaldsson, S.Þ. (1998). Seismotectonics of the central part of the South Iceland seismic zone. *Tectonophysics*, 298, 319–335.

Bergerat, F., Angelier, J., Verrier, S. (1999). Tectonic stress regimes, rift extension and transform motion: The South Iceland seismic zone. *Geodin. Acta*, 12(5), 303–319.

Bergerat, F., Angelier, J., Homberg, C. (2000). Tectonic analysis of the Husavik-Flatey Fault (northern Iceland) and mechanisms of an oceanic transform zone, the Tjörnes Fracture Zone. *Tectonics*, 19, 1161–1177.

Bergerat, F., Angelier, J., Guðmundsson, Á., Torfason, H. (2003). Push-ups, fracture patterns, and paleoseismology of the Leirubakki Fault, South Iceland. *J. Struct. Geol*, 25, 591–609.

Bergerat, F., Homberg, C., Angelier, J., Bellou, M. (2011). Surface traces of the Minnivellir, Réttarnes and Tjörvafit seismic faults in the South Iceland seismic zone: Segmentation, lengths and magnitude of related earthquakes. *Tectonophysics*, 498, 11–26.

Bergerat, F., Sæmundsson, K., Fourel, L., Angelier, J. (2013). Structure and activity of the geothermal field of Hvalfjörður (Iceland) from brittle tectonic, geothermal and paleostress analysis. In *Faults, Stresses and Mechanics of the Upper Crust: A Tribute to Jacques Angelier*, Lacombe, O., Bergerat, F. (eds). *Bull. Soc. Géol. France*, 184(4–5), 451–465.

Bjarnason, I.Þ., Cowie, P., Anders, M.H., Seeber, L., Scholz, C.H. (1993). The 1912 Iceland earthquake rupture: Growth and development of a nascent transform system. *Bull. Seismol. Soc. Am.*, 83, 416–435.

Björnsson, A. (1985). Dynamics of crustal rifting in NE Iceland. *J. Geophys. Res.*, 90(B12), 10.151–10.162.

Björnsson, G., Flóvenz, Ó.G., Sæmundsson, K., Einarsson, E. (2001). Pressure changes in Icelandic geothermal reservoirs associated with two large earthquakes in June 2000. *26th Workshop Geothermal Reservoir Engineering*. Stanford University, Stanford.

Bourgeois, O. (2000). Processus d'extension lithosphérique en Islande. Interactions avec les calottes glaciaires quaternaires. PhD Thesis, Université de Rennes I, Rennes.

Bourgeois, O., Dauteuil, O., Van Vliet-Lanoë, B. (1998). Subglacial volcanism in Iceland: Tectonic implications. *Earth Planet. Sci. Lett.*, 164(1–2), 165–178.

Bourgeois, O., Dauteuil, O., Hallot, E. (2005). Rifting above a mantle plume: Structure and development of the Iceland Plateau. *Geodin. Acta*, 18(1), 1–22.

Böðvarsson, R., Rögnvaldsson, S.Þ., Jakobsdóttir, S.S., Slunga, R., Stefánsson, R. (1996). The SIL data acquisition and monitoring system. *Seism. Res. Lett.*, 67(5), 35–46.

Browne, R.L. (2003). Hydrothermal alteration in active geothermal fields. *Ann. Rev. Earth Planet. Sci.*, 6(1), 229–224.

Campbell, K.A., Guido, D.M., Gautret, P., Foucher, F., Ramboz, C., Westall, F. (2015). Geyserite in hot-spring siliceous sinter: Window on Earth's hottest terrestrial (paleo)environment and its extreme life. *Earth. Sci. Rev.*, 148, 44–64.

Childs, C., Watterson, J., Walsh, J.J. (1995). Fault overlap zones within developing normal fault systems. *J. Geol. Soc. London*, 152, 535–549.

Childs, C., Nicol, A., Walsh, J.J., Watterson, J. (1996). Growth of vertically segmented normal faults. *J. Struct. Geol.*, 18, 1389–1397.

Clifton, A.E. and Einarsson, P. (2005). Styles of surface ruptures accompanying the June 17 and 21, 2000 earthquakes in the South Iceland seismic zones. *Tectonophysics*, 396, 141–159.

Clifton, A.E. and Kattenhorn, S.A. (2006). Structural architecture of a highly oblique divergent plate boundary segment. *Tectonophysics*, 419, 27–40.

Clifton, A.E., Pagli, C., Jónsdóttir, J.F., Eythórsdóttir, K., Vogfjörð, K. (2003). Surface effects of triggered fault slip on Reykjanes Peninsula, SW Iceland. *Tectonophysics*, 369, 145–154.

Cowie, P.A. and Roberts, G.P. (2001). Constraining slip rates and spacing for active normal faults. *J. Struct. Geol.*, 23, 1901–1915.

Daniel, G., Marsan, D., Bouchon, M. (2008). Earthquake triggering in southern Iceland following the June 2000 Ms 6.6 doublet. *J. Geophys. Res.*, 113(B05310) [Online]. Available at: http://doi.org/10.1029/2007jb005107.

Dauteuil, O., Angelier, J., Bergerat, F., Verrier, S., Villemin, T. (2001). Deformation partitioning inside a fissure swarm of the northern Icelandic rift. *J. Struct. Geol.*, 23, 1359–1372.

Dauteuil, O., Bourgeois, O., Mauduit, T. (2002). Lithosphere strength controls oceanic transform zone structure: Insights from analogue models. *Geophys. J. Int.*, 150, 706–714.

Decriem, J., Árnadóttir, Þ., Hooper, A., Geirsson, H., Sigmundsson, F., Keiding, M., Ófeigsson, B.G., Hreinsdóttir, S., Einarsson, P., La Femina, P., Bennett, R.A. (2010). The 2008 May 29 earthquake doublet in SW Iceland. *Geophys. J. Int.*, 181, 1128–1146.

DeMets, C., Gordon, R.G., Argus, D.F., Stein, S. (1990). Current plate motions. *Geophys. J. Int.*, 101, 425–478.

DeMets, C., Gordon, R.G., Argus, D.F., Stein, S. (1994). Effect of recent revisions to the geomagnetic reversal time scale on estimates of current plate motions. *Geophys. Res. Lett.*, 21, 2191–2194.

DeMets, C., Gordon, R.G., Argus, D.F. (2010). Geologically current plate motions. *Geophys. J. Int.*, 181, 1–80.

Duffield, W.A. (1975). Structure and origin of the Koae fault system, Kilauea Volcano, Haiwaii. Document, United States Government Printing Office, Washington.

Dyment, J., Lin, J., Baker, E. (2007). Ridge-hotspot interactions. What mid-ocean ridges tell us about deep earth processes. *Oceanography*, 20(1), 102–115.

Einarsson, P. (1979). Seismicity and earthquake focal mechanisms along the mid-Atlantic plate boundary between Iceland and the Azores. *Tectonophysics*, 55, 127–153.

Einarsson, P. (1987). Compilation of earthquake fault plane solutions in the North Atlantic and Arctic oceans. In *Recent Plate Movements and Deformation*, Kasahara, K. (ed.). American Geophysical Union, Washington.

Einarsson, P. (1991). Earthquakes and present-day tectonism in Iceland. *Tectonophysics*, 189, 261–279.

Einarsson, P. (2010). Mapping of Holocene surface ruptures in the South Iceland seismic zone. *Jökull*, 60, 117–134.

Einarsson, P. and Brandsdóttir, B. (2000). Earthquakes in the Mýrdalsjökull area, Iceland, 1978–1985: Seasonal correlation and connection with volcanoes. *Jökull*, 49, 59–73.

Einarsson, P. and Eiriksson, J. (1982). Earthquake fractures in the districts Land and Rangárvellir in the South Iceland seismic zone. *Jökull*, 32, 113–120.

Einarsson, P. and Sæmundsson, K. (1987). Earthquake epicenters 1982–1985 and volcanic systems in Iceland. In *Í hlutarins eðli. Festschrift for Þorbjörn Sigurgeirsson*, Sigfússon, Þ.I. (ed.). Menningarsjóður, Reykjavík.

Einarsson, P., Björnsson, S., Foulger, G., Stefánsson, R., Skaftadóttir, Þ. (1981). Seismicity pattern in the South Iceland seismic zone. In *Earthquake Prediction: An International Review*, Simpson, D.W., Richards, P.G. (eds). AGU, Washington.

Einarsson, P., Khodayar, M., Clifton, A., Ofeigsson, B., Thorjarnason, S., Einarsson, B., Hjartardottir, A.R. (2005). A map of Holocene fault structures in the South Iceland seismic zone. EGU General Assembly. European Geosciences Union, Vienna.

Einarsson, P., Hjartardóttir, Á.R., Hreinsdóttir, S., Imsland, P. (2018). The structure of seismogenic strike-slip faults in the eastern part of the Reykjanes Peninsula Oblique Rift, SW Iceland. *J. Volcan. Geotherm. Res.* [Online]. Available at: http://doi.org/10.1016/j.jvolgeores.2018.04.029.

Eldholm, O. and Grue, K. (1994). North Atlantic volcanic margins: Dimensions and production rates. *J. Geophys. Res.*, 99(B2), 2955–2968.

Fjäder, K., Guðmundsson, Á., Forslund, T. (1994). Dikes, minor faults and mineral veins associated with a transform fault in North Iceland. *J. Struct. Geol.*, 16, 109–119.

Flovenz, O.G. and Sæmundsson, K. (1993). Heat flow and geothermal processes in Iceland. *Tectonophysics*, 225, 123–138.

Forslund, T. and Guðmundsson, Á. (1991). Crustal spreading due to dikes and faults in Southwest Iceland. *J. Struct. Geol.*, 13, 443–457.

Garcia, S. (2003). Implications d'un saut de rift et du fonctionnement d'une zone transformante sur les déformations du nord de l'Islande. Approches structurale, sismotectonique et radio-chronologique. PhD Thesis, Université Pierre et Marie Curie, Paris.

Garcia, S. and Dhont, D. (2005). Structural analysis of the Húsavík–Flatey transform fault and its relationships with the rift system in northern Iceland. *Geodin. Acta*, 18(1), 31–41.

Garcia, S., Angelier, J., Bergerat, F., Homberg, C. (2002). Tectonic analysis of an oceanic transform fault zone based on fault–slip data and earthquake focal mechanisms: The Husavik–Flatey Fault zone, Iceland. *Tectonophysics*, 344, 157–174.

Garcia, S., Arnaud, N., Angelier, J., Bergerat, F., Homberg, C. (2003). Rift jump process in Northern Iceland since 10 MA for $^{40}Ar/^{39}Ar$ geochronology. *Earth Planet Sci. Lett.*, 241, 529–544.

Garcia, S., Angelier, J., Bergerat, F., Homberg, C., Dauteuil, O. (2008). Influence of rift jump and excess loading on the structural evolution of Northern Iceland. *Tectonics*, 27 [Online]. Available at: http://doi.org/10.1029/2006TC002029.

Geirsson, H., Árnadóttir, Þ., Völksen, C., Jiang, W., Sturkell, E., Villemin, T., Einarsson, P., Sigmundsson, F., Stefánson, R. (2006). Current plate movements across the Mid-Atlantic Ridge determined from 5 years of continuous GPS measurements in Iceland. *J. Geophys. Res.*, 111 [Online]. Available at: http://doi.org/10.1029/2005JB003717.

Geirsson, H., Hreinsdóttir, S., Decriem, J., La Femina, P.C., Jónsson, S., Bennett, R.A., Metzger, S., Holland, A., Sturkell, E., Villemin, T., Völksen, C., Sigmundsson, F., Einarsson, P., Roberts, M.J., Sveinbjörnsson, H. (2010). Overview of results from continuous GPS observations in Iceland from 1995 to 2010. *Jökull*, 60, 3–22.

Georgsson, L.S., Sæmundsson, K., Hjartarson, H. (2005). Exploration and development of the Hveravellir geothermal field, N-Iceland. *World Geothermal Congress*. WGC, Antalya, Turkey.

Grant, J.V. and Kattenhorn, S.A. (2004). Evolution of vertical faults at an extensional plate boundary, southwest Iceland. *J. Struct. Geol.*, 26, 537–557.

Greenfield, T., White, R.S., Roecker, S. (2016). The magmatic plumbing system of the Askja central volcano, Iceland, as imaged by seismic tomography. *J. Geophys. Res: Soldid Earth*, 121, 7211–7229.

Guðmundsson, Á. (1980). The Vogar fissure swarm, Reykjanes peninsula, SW Iceland. *Jökull*, 30, 43–64.

Guðmundsson, Á. (1986). Mechanical aspects of postglacial volcanism and tectonics of the Reykjanes Peninsula, Southwest Iceland. *J. Geophys. Res.*, 91, 12711–12721.

Guðmundsson, Á. (1987). Tectonics of the Thingvellir Fissure Swarm, SW Iceland. *J. Struct. Geol.*, 9, 61–69.

Guðmundsson, Á. (1992). Formation and growth of normal faults at the divergent plate boundary in Iceland. *Terra Nova*, 4, 464–471.

Guðmundsson, Á. (1995a). Stress fields associated with oceanic transform faults. *Earth Planet. Sci. Lett.*, 136, 603–614.

Guðmundsson, Á. (1995b). Ocean–ridge discontinuities in Iceland. *J. Geol. Soc. London*, 152, 1011–1015.

Guðmundsson, Á. (2000a). Active fault zones and groundwater flow. *Geophys. Res. Lett.*, 27(18), 2993–2996.

Guðmundsson, Á. (2000b). Dynamics of volcanic systems in Iceland. Example of tectonism and volcanism at juxtaposed hot spot and mid–ocean ridge systems. *Ann. Rev. Earth Planet. Sci.*, 8, 107–140.

Guðmundsson, Á. (2003). Surface stresses associated with arrested dykes in rift zones. *Bull. Volc.*, 65, 606–619.

Guðmundsson, Á. (2006). How local stresses control magma chamber ruptures, dyke injections, and eruptions in composite volcanoes. *Earth Sci. Rev.*, 79, 1–31.

Guðmundsson, Á. (2007). Infrastructure and evolution of ocean-ridge discontinuities in Iceland. *J. Geodin.*, 43, 6–29.

Guðmundsson, Á. and Bäckström, K. (1991). Structure and development of the Sveinagja Graben, Northeast Iceland. *Tectonophysics*, 200, 111–125.

Guðmundsson, Á. and Brynjólfsson, S. (1993). Overlapping rift-zone segments and the evolution of the South Iceland seismic zone. *Geophys. Res. Lett.*, 18, 1903–1906.

Guðmundsson, Á. and Fjäder, K. (1995). Dikes, minor faults and mineral veins associated with a transform fault in North Iceland: Reply. *J. Struct. Geol.*, 17, 1633–1636.

Guðmundsson, Á., Brynjólfsson, S., Jónsson, M.T. (1993). Structural analysis of a transform fault–rift zone junction in north Iceland. *Tectonophysics*, 220, 205–221.

Guðmundsson, Á., Bergerat, F., Angelier, J. (1996). Off-rift and rift-zone paleostresses in Northwest Iceland. *Tectonophysics*, 255, 211–228.

Guðmundsson, Á., Fjeldskaar, I., Brenner, S.L. (2002). Propagation pathways and fluid transport in jointed and layered rocks in geothermal fields. *J. Vol. Geotherm. Res.*, 116, 257–278.

Guðmundsson, Á., Lecoeur, N., Mohajeri, N., Þordarson, T. (2014). Dike emplacement at Bardarbunga, Iceland, induces unusual stress changes, caldera deformation, and earthquakes. *Bull. Volc.*, 76(869) [Online]. Available at: http://doi.org/10.1007/s00445-014-0869-8.

Gunnlaugsson, E. and Ívarsson, G. (2010). Direct use of geothermal water for district heating in Reykjavík and other towns and communities in SW-Iceland. *World Geothermal Congress*. Bali, Indonesia.

Hackman, M.C., King, G.C.P., Bilham, R. (1990). The mechanics of the South Iceland seismic zone. *J. Geophys. Res.*, 95(B11), 17339–17351.

Halldórsson, P. (2005). Jardskjálftavirkni á Nordurlandi. Report, Veðurstofa Íslands [Online]. Available at: https://www.vedur.is/media/vedurstofan/utgafa/greinargerdir/2005/05021.pdf.

Harðason, B.S., Fitton, J.G., Ellam, R.M., Pringle, M.S. (1997). Rift relocation–Geochemical and geochronological investigation of palaeorift in northwest Iceland. *Earth Planet. Sci. Lett.*, 153, 181–96.

Hartley, M.E. and Þórðarson, T. (2013). The 1874–1876 volcano-tectonic episode at Askja, North Iceland: Lateral flow revisited. *Geochem. Geophys. Geosyst.*, 14, 2286–2309.

Hasegawa, H.S. and Basham, P. (1989). Spatial correlation between seismicity and postglacial rebound in eastern Canada. In *Earthquakes at North–Atlantic Passive Margins: Neotectonics and Postglacial Rebound*, Gregerson, S., Basham, P. (eds). Kluwer, Dordecht.

Helgason, J. and Zentilli, M. (1985). Fields characteristics of laterally emplaced dykes: Anatomy of an exhumed Miocene dike swarm in Reydarfjördur, Eastern Iceland. *Tectonophysics*, 115, 247–274.

Hjaltadóttir, S. (2009). Use of relatively located microearthquakes to map fault patterns and estimate the thickness of the brittle crust in Southwest Iceland. Thesis, Reykjavík University, Reykjavík.

Hjartardóttir, Á.R., Einarsson, P., Sigurdsson, H. (2009). The fissure swarm of the Askja volcanic system along the divergent plate boundary of N Iceland. *Bull. Volc.*, 71, 961–975.

Hjartardóttir, Á.R., Einarsson, P., Guðmundsson, M.T., Högnadóttir, T. (2016a). Fracture movements and graben subsidence during the 2014 Bárðarbunga dike intrusion in Iceland. *J. Volcan. Geother. Res.*, 310, 242–252.

Hjartardóttir, Á.R., Einarsson, P., Magnúsdóttir, S., Björnsdóttir, Þ., Brandsdóttir, B. (2016b). Fracture systems of the Northern Volcanic Rift Zone, Iceland: An onshore part of the mid-Atlantic plate boundary. In *Magmatic Rifting and Active Volcanism*, Wright, T.J., Ayele, A., Ferguson, D.J., Kidane, T., Vye-Brown, C. (eds). Geological Society of London, London.

Hjartarson, Á., Erlendsson, Ö., Blischke, A. (2017). The Greenland–Iceland–Faroe Ridge complex. In *The NE Atlantic Region: A Reappraisal of Crustal Structure, Tectonostratigraphy and Magmatic Evolution*, Péron-Pinvidic, G., Hopper, J.R., Stoker, M., Gaina, C., Funck, T., Árting, U.E., Doornenbal, J.C. (eds). Geological Society of London, London.

Homberg, C., Bergerat, F., Angelier, J., Garcia, S. (2010). Fault interaction and stresses along broad oceanic transform zone: Tjörnes fracture zone, north Iceland. *Tectonics*, 29 [Online]. Available at: http://doi.org/10.1029/2008TC002415.

Hreinsdóttir, S., Sigmundsson, F., Roberts, M.J., Björnsson, H., Grapenthin, R., Arason, P., Árnadóttir, Þ., Hólmjárn, J., Geirsson, H., Bennett, R.A., Guðmundsson, M.T., Oddsson, B., Ófeigsson, B.G., Villemin, T., Jónsson, Þ., Sturkell, E., Höskuldsson, Á., Larsen, G., Þordarson, Þ., Óladóttir, B.A. (2014). Volcanic plume height correlated with magma pressure change at Grímsvötn Volcano, Iceland. *Nat. Geosci.*, 7(3), 214–218.

Huggins, P., Watterson, J., Walsh, J.J., Childs, C. (1995). Relay zone geometry and displacement transfer between normal faults recorded in coal-mine plans. *J. Struct. Geol.*, 17, 1741–1755.

Jakobsdóttir, S.S. (2008). Seismicity in Iceland: 1994–2007. *Jökull*, 58, 75–100.

Jakobsdóttir, S.S., Guðmundsson, G.B., Stefánsson, R. (2002). Seismicity in Iceland 1991–2000 monitored by the SIL seismic system. *Jökull*, 51, 87–94.

Jakobsdóttir, S.S., Roberts, M.J., Guðmundsson, G.B., Geirsson, H., Slunga, R. (2008). Earthquake swarms at Upptyppingar, north–east Iceland: A sign of magma intrusion? *Stud. Geophys. Geod.*, 52, 513–528.

Jakobsson, S.P. (1979). Outline of the petrology of Iceland. *Jökull*, 29, 57–73.

Jancin, M., Young, K.D., Voight, B. (1985). Stratigraphy and K/Ar ages across the west flank of the Northeast Iceland axial rift zone, in relation to the 7 Ma volcanotectonic reorganisation of Iceland. *J. Geophys. Res.*, 90, 9961–9985.

Jancin, M., Young, K.D., Voight, B., Orkan, N.I. (1995). Dikes, minor faults and mineral veins associated with the transform fault in North Iceland: Discussion. *J. Struct. Geol.*, 17, 1627–1631.

Jouanne, F., Villemin, T., Berger, A., Henriot, O. (2006). Rift-transform junction in North Iceland: Rigid blocks and narrow accommodation zones revealed by GPS 1997–1999–2002. *Geophys. J. Intern.*, 167, 1439–1446.

Jóhannesson, H. and Sædmundsson, K. (1998). Carte géologique de l'Islande à 1/500 000, 2nd edition. Document, Icelandic Institute of Natural History, Reykjavík.

Jóhannesson, H., Jakobsson, S.P., Sædmundsson, K. (1990). Carte géologique de l'Islande à 1/250 000, feuille 6 : sud–Islande, 3rd edition. Document, Icelandic Museum of Natural History and Iceland Geodetic Survey, Reykjavík.

Jónsdóttir, K., Tryggvason, A., Roberts, R., Soosalu, H., Böðvarsson, R. (2007). Habits of a glacier covered volcano: Seismicity and structure study of the Katla volcano, south Iceland. *Ann. Glaciol.*, 45, 169–177.

Jónsdóttir, K., Roberts, R., Pohjola, V., Lund, B., Shomali, Z.H., Tryggvason, A., Böðvarsson, R. (2009). Glacial long period seismic events at Katla volcano, Iceland. *Geophys. Res. Lett.*, 36, 200 [Online]. Available at: http://doi.org/10.1029/2009GL038234.

Jónsson, S. (2008). Multi-year postseismic ground movements in south Iceland dominated by viscous relaxation. *Nat. Geosci.*, 1, 136–139.

Karnik, V. (1969). *Seismicity of the European Area? Volume I.* Reidel, Dordrecht.

Karson, J.A., Farrell, J.A., Chutas, L.A., Nanfito, A.F., Proett, J.A., Runnals, K.T., Sæmundsson, K. (2018). Rift-parallel strike-slip faulting near the Iceland plate boundary zone: Implications for propagating rifts. *Tectonics*, 37, 4567–4594 [Online]. Available at: http://doi.org/10.1029/2018TC005206.

Kim, Y.S., Sanderson, D.J. (2005). The relationship between displacement and length of faults: A review. *Earth Sci. Rev.*, 68, 317–334.

La Femina, P.C., Dixon, T.H., Malservisi, R., Arnardóttir, T., Sturkell, E., Sigmundsson, F., Einarsson, P. (2005). Strain partitioning and accumulation in a propagating ridge system: Geodetic GPS measurements in south Iceland. *J. Geophys. Res.*, 110(B11405) [Online]. Available at: http://doi.org/10.1029/2005JB003675.

Långbacka, B.O. and Guðmundsson, Á. (1995). Extensional tectonics in the vicinity of a transform fault in North Iceland. *Tectonics*, 14, 294–306.

Le Breton, E., Cobbold, P.R., Dauteuil, O., Lewis, G. (2012). Variations in amount and direction of seafloor spreading along the northeast Atlantic Ocean and resulting deformation of the continental margin of northwest Europe. *Tectonics*, 31 [Online]. Available at: http://doi.org/10.1029/2011TC003087.

Leifsson, M. (1992). Industrial use of the high–temperature geothermal field at Theistareykir, Iceland. *Geothermics*, 21(5–6), 631–640.

MacDonald, G.A. (1957). Faults and monoclines on Kilauea Volcano, Hawaii. *Geol. Soc. Amer. Bull.*, 68, 269–271.

Mamula, N. and Voight, B. (1985). Tectonic analysis of lineaments near a spreading axis, northeastern Iceland. *Tectonophysics*, 116, 63–93.

Mansfield, C.S. and Cartwright, J.A. (1996). High resolution fault displacement mapping from three-dimensional seismic data: Evidence for dip linkage during fault growth. *J. Struct. Geol.*, 18, 249–263.

Martel, S.J. and Langley, J.S. (2006). Propagation of normal faults to the surface in basalt, Koae fault system, Hawaii. *J. Struct. Geol.*, 28, 2123–2143.

McDougall, I. and Wensink, H. (1966). Paleomagnetism and geochronology of the Pliocene Pleistocene lavas in Iceland. *Earth Planet. Sci. Lett.*, 1, 232–236.

McDougall, I., Watkins, N.D., Kristjansson, L. (1976). Geochronology and paleomagnetism of a Miocene–Pliocene lava sequence at Bessastadaa, eastern Iceland. *Am. J. Sci.*, 276, 1078–1095.

McDougall, I., Sæmundsson, K., Jóhannesson, H., Watkins, N.D., Kristjansson, L. (1977). Extension of the geomagnetic polarity time scale to 6.5 m.y.: K-Ar dating, geological and paleomagnetic study of a 3,500-m lava succession in western Iceland. *Geol. Soc. Am. Bull.*, 88, 1–15.

McDougall, I., Kristjansson, L., Sæmundsson, K. (1984). Magnetostratigraphy and geochronology of NW Iceland. *J. Geophys. Res.*, 89(B8), 7029–7060.

Metzger, S., Jónsson, S., Geirsson, H. (2011). Locking depth and slip-rate of the Húsavík Flatey fault, North Iceland, derived from continuous GPS data 2006–2010. *Geophys. J. Int.*, 187(2), 564–576.

Metzger, S., Jónsson, S., Danielsen, G., Hreinsdóttir, S., Jouanne, F., Giardini, D., Villemin, T. (2013). Present kinematics of the Tjörnes fracture zone, North Iceland, from campaign and continuous GPS measurements. *Geophys. J. Int.*, 192, 441–455.

Meyer, P.S., Sigurdsson, H., Schilling, J.G. (1985) Petrological and geochemical variations along Iceland's neovolcanic zones. *J. Geophys. Res.*, 90, 10043–10072.

Moorbath, S., Sigurdsson, H., Goodwin, R. (1968). K-Ar ages of the oldest exposed rocks in Iceland. *Earth Planet. Sci. Lett.*, 4, 197–205.

Musset, A.E., Ross, J.G., Gibson, I.L. (1980). $^{40}Ar/^{39}Ar$ dates of eastern Iceland lavas. *R. Astron. Soc. Geophys. J.*, 60, 37–52.

Neale, C.M.U., Jaworowski, C., Heasler, H., Sivarajan, S., Masih, A. (2016). Hydrothermal monitoring in Yellowstone National Park using airborne thermal infrared remote sensing. *Remote Sens. Environ.*, 184, 628–644.

Opheim, J.A. and Guðmundsson, Á. (1989). Formation and geometry of fractures and related volcanism of the Krafla fissure swarm, northeast Iceland. *Geol. Soc. Am. Bull.*, 101, 1608–1622.

Pagli, C., Pedersen, R., Sigmundsson, F., Feigl, K.L. (2003). Triggered fault slip on June 17, 2000 on the Reykjanes Peninsula, SW–Iceland captured by radar interferometry. *Geophys. Res. Lett.*, 30(6), 1273–1276.

Pálmason, G. (1981). Crustal rifting, and related thermomechanical processes in the lithosphere beneath Iceland. *Geol. Rundsch.*, 70, 244–260.

Papanikolaou, I.D. and Roberts, G.P. (2007). Geometry, kinematics and deformation rates along the active normal fault system in the southern Apennines: Implications for fault growth. *J. Struct. Geol.*, 29, 166–188.

Paquet, F., Dauteuil, O., Hallot, E., Moreau, F. (2007). Tectonics and magma dynamics coupling in a dyke swarm of Iceland. *J. Struct. Geol.*, 29, 1477–1493.

Peacock, D.C.P. (1991). Displacements and segment linkage in strike-slip fault zones. *J. Struct. Geol.*, 13, 721–733.

Peacock, D.C.P. and Parfitt, E.A. (2002). Active relay ramps and normal fault propagation on Kilauea Volcano, Hawaii. *J. Struct. Geol.*, 24, 729–742.

Peacock, D.C.P. and Zhang, X. (1993). Field examples and numerical modeling of oversteps and bends along normal faults in cross-section. *Tectonophysics*, 234, 147–167.

Pedersen, R., Sigmundsson, F., Feigl, K.L., Árnadóttir, Þ. (2001). Co-seismic interferograms of two Ms = 6.6 earthquakes in the South Iceland seismic zone, June 2000. *Geophys. Res. Lett.*, 28, 3341–3344.

Pedersen, R., Jonsson, S., Árnadóttir, Þ., Sigmundsson, F., Feigl, K. (2003). Fault slip distribution of two June 2000 M(w) 6.5 earthquakes in South Iceland estimated from joint inversion of InSAR and GPS measurements. *Earth Planet. Sci. Lett.*, 213, 487–502.

Pedersen, R., Sigmundsson, F., Einarsson, P. (2007). Controlling factors on earthquake swarms associated with magmatic intrusions. Constraints from Iceland. *J. Volcan. Geotherm. Res.*, 162, 73–80.

Philipp, S.L., Guðmundsson, Á., Oelrich, A.R.I. (2007). How structural geology can contribute to make geothermal projects successful? *European Geothermal Congress.* Unterhaching, Germany.

Plateaux, R. (2012). Architecture et mécanismes du rift islandais dans la région du Vatnajökull. PhD Thesis, Université de Nice-Sophia-Antipolis, Nice, France.

Plateaux, R., Bergerat, F., Béthoux, N., Villemin, T., Gerbault, M. (2012). Implications of fracturing mechanisms and fluid overpressure in the east Iceland rift zone, from earthquakes and fault slip data. *Tectonophysics*, 581, 19–34.

Ragnarsson, Á. (2008). Utilization of geothermal energy in Iceland. *14th Building Services.* Mechanical and Building Industry Days, Debrecen, Hungary.

Reverso, T., Vandemeulebrouck, J., Jouanne, F., Pinel, V., Villemin, T., Sturkell, E., Bascou, P. (2014). A two-magma chamber model as a source of deformation at Grimsvötn Volcano, Iceland. *J. Geophys. Res.*, 119, 4666–4683.

Riedel, W. (1929). Zur Mechanik geologischer Brucherscheinungen. *Zentralblatt für Mineralogie, Geologie und Paläontologie*, 1929B, 354–368.

Riedel, C., Tryggvason, A., Dahm, T., Stefánsson, R., Böðvarsson, R., Guðmundsson, G.B. (2005). The seismic velocity structure north of Iceland from joint inversion of local earthquake data. *J. Seismol.*, 9, 383–404.

Roberts, G.P. and Michetti, A.M. (2004). Spatial and temporal variations in growth rates along active normal fault systems: An example from The Lazio-Abruzzo Apennines, central Italy. *J. Struct. Geol.*, 26, 339–376.

Roth, F. (2004). Stress changes modelled for the sequence of strong earthquakes in the South Iceland seismic zone since 1706. *Pure appl. Geophys.*, 161(7), 1305–1327.

Rögnvaldsson, S.T. and Slunga, R. (1994). Single and joint fault plane solutions for micro-earthquakes in South Iceland. *Tectonophysics*, 237, 73–86.

Rögnvaldsson, S.T., Guðmundsson, Á., Slunga, R. (1998). Seismotectonic analysis of the Tjörnes fracture zone, an active transform fault in north Iceland. *J. Geophy. Res.*, 103(B12), 30117–30129.

Rowland, J.V., Baker, E., Ebinger, C.J., Keir, D., Kidane, T., Biggs, J., Hayward, N.T., Wright, T.J. (2007). Fault growth at a nascent slow-spreading ridge: 2005 Dabbahu rifting episode, Afar. *Geophys. J. Int.*, 171, 1226–1246.

Sæmundsson, K. (1974). Evolution of the axial rifting zone in northern Iceland and the Tjörnes fracture zone. *Geol. Soc. Am. Bull.*, 85, 495–504.

Sæmundsson, K. (1978). Fissure swarms and central volcanoes of the neovolcanic zones of Iceland. *Geol. J.*, 10, 415–432.

Sæmundsson, K. (1979). Outline of the geology of Iceland. *Jökull*, 29, 7–28.

Sæmundsson, K. (1991). Jarfraedi Kroflukerfisins. In *Nattura Myvatns*, Gardarsson, A., Einarsson, P. (eds). Hid Islenska Natturufraedifelag, Reykjavík.

Sæmundsson, K. (2011). Dulinn lághiti og virk eldstöðvakerfi (Hidden low temperature systems and active volcanic systems) [in Icelandic]. Document, Conférence Com. Géol. d'Islande, 29–32.

Sæmundsson, K. (2014). Breytt mynd af hitastigli (Modified image of the temperature gradient) [in Icelandic]. Document, Conférence Com. Géol. d'Islande, 27.

Sæmundsson, K. and Einarsson, S. (1980). Carte géologique de l'Islande à 1/250 000, feuille 3: Sud–Ouest Islande, 2nd edition. Document, Museum of Natural History and Iceland Geodetic Survey, Reykjavík.

Sæmundsson, K., Kristjansson, L., McDougall, I., Watkins, N.D. (1980). K-Ar dating, geological and paleomagnetic study of a 5-km lava succession in northern Iceland. *J. Geophys. Res.*, 85(B7), 3628–3646.

Sæmundsson, K., Axelsson, G., Steingrímsson, B. (2009). Geothermal systems in global perspective. *Short Course on Surface Exploration for Geothermal Resources*. UNU-GTP-LaGeo, Ahuachapan-Santa Tecla, Salvador.

Sæmundsson, K., Sigurgeirsson, M.Á., Friðleifsson, G.Ó. (2018). Geology and structure of the Reykjanes volcanic system, Iceland. *J. Volcan. Geoth. Res.* [Online]. Available at: http://doi.org/10.1016/j.jvolgeores.2018.11.022.

Schöpfer, M.P.J., Childs, C., Walsh, J.J. (2006). Localisation of normal faults in multilayer sequences. *J. Struct. Geol.*, 28, 816–833.

Sclater, J.G., Jaupart, C., Galson, D. (1980). The heat flow through oceanic and continental crust and the heat loss of the Earth. *Rev. Geophys. Space Phys.*, 18, 269–311.

Sgattoni, G., Guðmundsson, Ó., Einarsson, P., Lucchi, F., Li, K.L., Sadeghisorkhani, H., Roberts, R., Tryggvason, A. (2017). The 2011 unrest at Katla volcano: Characterization and interpretation of the tremor sources. *J. Volcan. Geotherm. Res.*, 338, 63–78.

Sigmundsson, F. (2006). *Iceland Geodynamics. Crustal Deformation and Divergent Plate Tectonics*. Springer, Chichester.

Sigmundsson, F., Einarsson, P., Bilham, R., Sturkell, E. (1995). Rift-transform kinematics in south Iceland: Deformation from Global Positioning System measurements, 1986–1992. *J. Geophys. Res.*, 100, 6235–6248.

Sigmundsson, F., Hooper, A., Hreinsdóttir, S., Vogfjörð, K.S., Ofeigsson, B.G., Heimisson, E.R., Dumont, S., Parks, M., Spaan, K., Guðmundsson, G.B., Drouin, V., Árnadóttir, Þ., Jonsdóttir, K., Guðmundsson, M.T., Hognadóttir, T., Fridriksdóttir, H.M., Hensch, M., Einarsson, P., Magnusson, E., Samsonov, S., Brandsdóttir B., White, R.S., Ágústsdóttir, T., Greenfield, T., Green, R.G., Hjartardóttir, A.R., Pedersen, R., Bennet, R.A., Geirsson, H., La Femina, P.C., Bjornsson, H., Palsson, F., Sturkell, E., Bean, C.J., Mollhoff, M., Braiden, A.K., Eibl, E.P.S. (2015). Segmented lateral dyke growth in a rifting event at Bárðarbunga volcanic system, Iceland. *Nature*, 517, 191–195.

Sigmundsson, F., Einarsson, P., Hjartardóttir, Á.R., Drouin, V., Jónsdóttir, K., Árnadóttir, Þ., Geirsson, H., Hreinsdóttir, S., Li, S., Ófeigsson, B.G. (2018a). Geodynamics of Iceland and the signatures of plate spreading. *J. Volcan. Geotherm. Res.* [Online]. Available at: http://doi.org/10.1016/j.jvolgeores.2018.08.014.

Sigmundsson, F., Parks, M., Pedersen, R., Jónsdóttir, K., Ófgeisson, B.G., Grapenthin, R., Dumont, S., Einarsson, P., Drouin, V., Heimisson, E.R., Hjartardóttir, A.R., Guðmundsson, M., Geirsson, H., Hreinsdóttir, S., Sturkell, E., Hooper, A., Högnadóttir, Þ., Vogfjörð, K., Roberts M.J. (2018b). Magma movements in volcanic plumbing systems and their associated ground deformation and seismic patterns. In *Volcanic and Igneous Plumbing Systems*, Burchard, S. (ed.). Elsevier, Cambridge.

Sinton, J., Grönvold, K., Sæmundsson, K. (2005). Postglacial eruptive history of the Western Volcanic Zone, Iceland. *Geochem. Geophys. Geosyst.*, 6 [Online]. Available at: http://doi.org/10.1029/2005GC001021.

Smith, P.J. (1967). The intensity of the Tertiary geomagnetic field. *Geophys. J. Roy. Astron. Soc.*, 12, 239–258.

Sonnette, L., Angelier, J., Villemint, T., Bergerat, F. (2010). Faulting and fissuring in active oceanic rift: Surface expression, distribution and tectonic–volcanic interaction in the Thingvellir Fissure Swarm, Iceland. *J. Struct. Geol.*, 32, 407–422.

Soosalu, H., Einarsson, P., Þorbjarnardóttir, B.S. (2005). Seismic activity related to the 2000 eruption of the Hekla volcano, Iceland. *Bull. Volc.*, 68, 21–36.

Soosalu, H., Key, A.J., White, R.S., Knox, C., Einarsson, P., Jakobsdóttir, S.S. (2010). Lower crustal earthquakes caused by magma movement beneath Askja volcano on the north Iceland rift. *Bull. Volc.*, 72, 55–62.

Stefánsson, R. (1966). Methods of focal mechanism studies with application to two Atlantic earthquakes. *Tectonophysics*, 3, 209–243.

Stefánsson, R. (2006). PREPARED–Final report. Report, Veðurstofa Íslands [Online]. Available at: https://www.vedur.is/media/vedurstofan/utgafa/greinargerdir/2006/06009.pdf [Accessed 30 August 2019].

Stefánsson, R. and Halldórsson, P. (1988). Strain release and strain build–up. *Tectonophysics*, 152, 267–276.

Stefánsson, R., Bödvardsson, R., Slunga, R., Einarsson, P., Jakobsdóttir, H., Bungum, S., Gregersen, S., Havskov, J., Hjelme, J., Korhonen, H. (1993). Earthquake prediction research in the South Iceland seismic zone and the SIL project. *Bull. Seism. Soc. Am.*, 83, 696–716.

Stefánsson, R., Guðmundsson, G.B., Halldórsson, P. (2000). The two large earthquakes in the South Iceland seismic zone on June 17 and 21, 2000. Report, Veðurstofa Íslands [Online]. Available at: http://hraun.vedur.is/ja/skyrslur/June17and21_2000/index.html [Accessed 30 August 2019].

Stefánsson, R., Bonafede, M., Roth, F., Einarsson, P., Árnadóttir, Þ., Guðmundsson, G.B. (2006). Modelling and parameterizing the SW Iceland earthquake release and deformation process. Report, Veðurstofa Íslands [Online]. Available at: https://www.vedur.is/media/vedurstofan/utgafa/greinargerdir/2006/06005.pdf.

Stefánsson, R., Guðmundsson, G.B., Halldorsson, P. (2008). Tjörnes fracture zone. New and old seismic evidences for the link between the North Iceland rift zone and the Mid-Atlantic ridge. *Tectonophysics*, 447, 117–126.

Steinþórsson, S.N., Óskarsson, N., Sigvaldason, G.E. (1985). Origin of alkali basalts in Iceland: A plate tectonic model. *J. Geophys. Res.*, 90(B12), 10027–10042.

Stetter, K.O. (1996). Hyperthermophilic procaryotes. *FEMS Microbiol. Rev.*, 18, 149–158.

Stirling, M., Rhoades, D., Berryman, K. (2002). Comparison of earthquake scaling relations derived from data of the instrumental and preinstrumental era. *Bull. Seism. Soc. Am.*, 92(2), 812–830.

Sturkell, E., Einarsson, P., Sigmundsson, F., Geirsson, H., Pedersen, R., Van Dalfsen, E., Linde, A., Sacks, S., Stefánsson, R. (2006). Volcano geodesy and magma dynamics in Iceland. *J. Volcan. Geotherm. Res.*, 150, 14–34.

Sudhaus, H. and Jónsson, S. (2008). Improved source modelling through combined use of InSAR and GPS under consideration of correlated data errors: Application to the June 2000 Kleifarvatn earthquake, Iceland. *Geophys. J. Int.*, 176, 389–404.

Takacs, C.D., Ehringer, M., Favre, R., Cermola, M., Eggertsson, G., Pálsdóttir, A., Reysenbach, A.L. (2001). Phylogenetic characterization of the blue filamentous bacterial community from an Icelandic geothermal spring. *FEMS Microbiol. Ecol.*, 35, 123–128.

Tentler, T. (2005). Propagation of brittle failure triggered by magma in Iceland. *Tectonophysics*, 406, 17–38.

Thoroddsen, Þ. (1899). *Jardskalftar a Surdulandi*. Société littéraire islandaise, Copenhagen.

Thoroddsen, Þ. (1905). *Landskjálftar á Íslandi*. Société littéraire islandaise, Copenhagen.

Tobler, D.J., Stefánsson, A., Benning, L.G. (2008). In-situ grown silica sinters in Icelandic geothermal areas. *Geobiol.*, 6, 481–502.

Tómasson, J. and Kristmannsdóttir, H. (1972). High temperature alteration minerals and thermal brines, Reykjanes, Iceland. *Contr. Mineral. Petrol.*, 36(2), 123–134.

Travis, M.E. (2012). Inter-rifting and inter-seismic strain accumulation in a propagating ridge system: A study from south Iceland. Master's Thesis, Pennsylvania State University, State College, PA.

Tryggvason, E. (1973). Seismicity, earthquake swarms and plate boundaries in the Iceland region. *Bull. Seismol. Soc. Amer.*, 63(4), 1327–1348.

Tryggvason, E., Husebye, E.S., Stefánsson, R. (1983). Seismic image of the hypothesized Icelandic hot spot. *Tectonophysics*, 100, 97–118.

Van Vliet-Lanoë, B., Bourgeois, O., Dauteuil, O., Embry, J.C., Guillou, H., Schneider, J.L. (2005). Deglaciation and volcano-seismic activity in Northern Iceland: Holocene and early Eemian (The Syðra Formation). *Geodin. Acta*, 18, 81–100.

Van Vliet-Lanoë, B., Bergerat, F., Allemand, P., Innocent, C., Guillou, H., Cavailhes, T., Liorzou, C., Grandjean, P., Passot, S. (2020a). Tectonism and volcanism enhanced by deglaciation events in southern Iceland. *Quat. Res.*, 94, 94–120 [Online]. Available at: http://doi.org/10.1017/QUA.2019.68.

Van Vliet-Lanoë, B., Knudsen, O., Guðmundsson, A., Guillou, H., Chazot, G., Langlade, J., Liorzou, C., Nonnotte, N. (2020b). Volcanoes and climate: The triggering of Preboreal jökulhlaups in Iceland. *Int. J. Earth Sci.*, 109, 847–876 [Online]. Available at: http://doi.org/10.1007/s00531-020-01833-9.

Villemin, T. and Bergerat, F. (2013). From surface fault traces to a fault growth model: The Vogar fissure swarm of the Reykjanes Peninsula, Southwest Iceland. *J. Struct. Geol.*, 51, 38–51.

Vogfjörð, K.S., Jakobsdóttir, S.S., Guðmundsson, G.B., Roberts, M.J., Ágústsson, K., Arason, T., Geirsson, H., Karlsdóttir, S., Hjaltadóttir, S., Ólafsdóttir, U., Thorbjarnardóttir, B., Skaftadóttir, T., Sturkell, E., Jónasdóttir, E.B., Hafsteinsson, G., Sveinbjörnsson, H., Stefánsson, R., Jónsson, T.V. (2005). Forecasting and monitoring of a volcanic eruption in Iceland. *Eos*, 86(26) [Online]. Available at: http://doi.org/10.1029/2005 EO260001.

Walter, T.R., Jousset, P., Allahbakhshi, M., Witt, T., Guðmundsson, M.T., Hersir, G.P. (2018). Underwater and drone based photogrammetry reveals structural control at Geysir geo-thermal field in Iceland. *J. Volcan. Geotherm. Res.*, 391 [Online]. Available at: http://doi.org/10.1016/j.jvolgeores.2018.01.010.

Walters, R.L., Jones, S.M., Maclennan, J. (2013). Renewed melting at the abandoned Húnaflói Rift, northern Iceland, caused by plume pulsing. *Earth Planet. Sci. Lett.*, 377–378, 227–238.

Ward, P.L. (1971). New interpretation of the geology of Iceland. *Geol. Soc. Am. Bull.*, 82, 2991–3012.

Young, K.D., Jancin, M., Voight, B., Orkan, N.I. (1985). Transform deformation of Tertiary rocks along the Tjörnes fracture zone. *J. Geophys. Res.*, 90, 9986–10010.

Young, K.D., Orkan, N., Jancin, M., Sæmundsson, K., Voight, B. (2020). Major tectonic rotation along an oceanic transform zone, northern Iceland: Evidence from field and paleomagnetic investigations. *J. Volcan. Geotherm. Res.*, 391 [Online]. Available at: http://doi.org/10.1016/j.jvolgeores.2018.11.020.

Þórarinsson, S. (1937). Das Dalvik–Beben in Nordisland. *Geogr. Ann.*, 19, 232–277.

Þórarinsson, S. and Sæmundsson, K. (1979). Volcanic activity in historical time. *Jökull*, 29, 29–32.

Þorbjarnardóttir, B.S. and Guðmundsson, G.B. (2003). Seismicity in Iceland 2002. *Jökull*, 53, 49–54.

Þórðarson, T. and Larsen, G. (2007). Volcanism in Iceland in historical time: Volcano types, eruption styles and eruptive history. *J. Geodynamics*, 43, 118–152.

3

Iceland, a Legacy of North Atlantic History

Laurent GEOFFROY with the collaboration of René MAURY and Brigitte VAN VLIET-LANOË

Understanding the beginnings and significance of Iceland requires taking into account both the present-day characteristics of the Northeast Atlantic and the geodynamic history of this region since the Paleozoic. The history of the opening of the Northeast Atlantic indeed testifies to the complexity of the interactions between the dynamics of a partially molten sub-lithospheric mantle and that of the tectonic fragmentation of a continent deeply marked by its geodynamic evolution since the Caledonian.

3.1. Bathymetry of the Northeast Atlantic domain and geoid anomalies

Remarkably, basin bathymetry as well as crustal floor depths in the Labrador–Baffin axis and the Northeast Atlantic are unusually low (Figure 3.1(a)).

The oceanic lithosphere continuously formed on either side of the oceanic expansion axes is progressively cooling and, therefore, thickening. This process leads to a subsidence of this lithosphere, which becomes denser due to thermal contraction. It results in a deepening of the oceanic bedrock roof under the marine sediments, increasing with the age of the oceanic lithosphere.

For color versions of the figures in this chapter see www.iste.co.uk/vanvliet/iceland1.zip.

Iceland Within the Northern Atlantic 1,
coordinated by Brigitte VAN VLIET-LANOË.

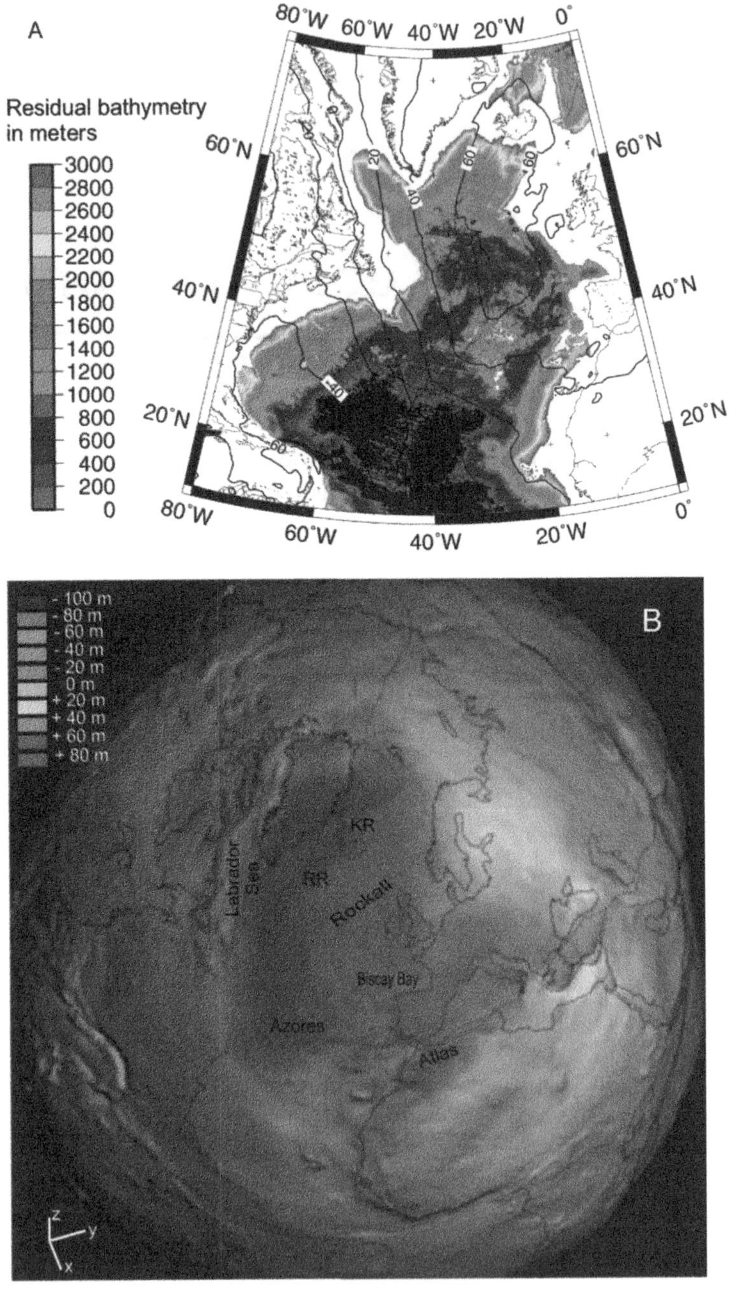

Figure 3.1. *Residual topography of the North Atlantic and Geoid*

COMMENT ON FIGURE 3.1.– *(A) The values correspond to the difference between the theoretical thermal subsidence bathymetry of the ocean lithosphere and the actual bathymetry. The curves correspond to the values in m of the geoid anomaly represented in B (King 2005). (B) Geoid anomalies in the Azores-Northeast Atlantic region, over Western Europe and North Africa (completed from (Köhler 2004)). RR: Reykjanes Ridge. KR: Kolbeinsey Ridge.*

The oldest oceanic lithosphere in the Northeast Atlantic system, south of the Labrador Sea, is dated at more than 65 My and should therefore be associated with a depth of about 4,500 m. This depth is only reached south of the Charlie–Gibbs transform fault (latitude 53° N, Figures 1.1 and 1.5), i.e. outside the Northeast Atlantic system.

Globally, a deficit in the depth of the seafloor of 1 to more than 3 km is observed in the entire Northeast Atlantic (Figure 3.1(a)). This deficit is very pronounced in the Davis Strait between the Labrador Sea and Baffin Bay, as well as in the GIFR where particularly low bathymetry is encountered, in the order of –200 to –500 m. The highest point of the bathymetric anomaly is Iceland, with an average altitude of +500 m, while the mean depth of the oceanic ridges worldwide is close to –2,500 m. Iceland is thus the only known subaerial active ridge.

These bathymetric anomalies are correlated with a very significant regional geoid anomaly (Cazenave and Feigl 1994) (Figure 3.1(b)). Globally, one of the most significant positive geoid anomalies is that of the North Atlantic zone with a maximum of +75 m (Figure 3.1(b)). This maximum is centered south of Iceland. This geoid anomaly zone is significantly larger than the Northeast Atlantic bathymetric anomaly (Figure 3.1), but the bathymetric anomaly zone is perfectly included within the geoid anomaly. This long wavelength geoid anomaly, which reflects a regional excess mass, is generally interpreted as related to a lithospheric uplift. The latter would follow a dynamic upwelling of a relatively warm and, therefore, low viscous asthenosphere (Cazenave *et al.* 1989; King 2005).

3.2. The North Atlantic and the continental breakup of Laurussia

The progressive splitting of a supercontinent, Laurussia (Figure 3.2), led to the opening of the northeast Atlantic. Laurussia became individualized during the Silurian (Paleozoic) from the coalescence by continental collision (Caledonian orogeny) of three landmasses, Laurentia (including North America and Greenland), Baltica and Avalonia.

One of the fundamental questions in geodynamics is to understand how a continent, rheologically multilayered, can stretch and thin out until it breaks mechanically up. In this situation, a convective regime is established, giving rise to a new ocean lithosphere. It is the study of "passive margins" that makes it possible to decipher these processes of continental rupture. These margins are the transition zones between continental and oceanic lithospheres, which appear during the breakup of a continental lithosphere after stretching and thinning. They present distinct structures according to the amount of magmatism associated with the extension (Figure 3.3). A reminder of the structure of the passive margins is essential to understand the breakup of Laurussia and the possible origin of Iceland.

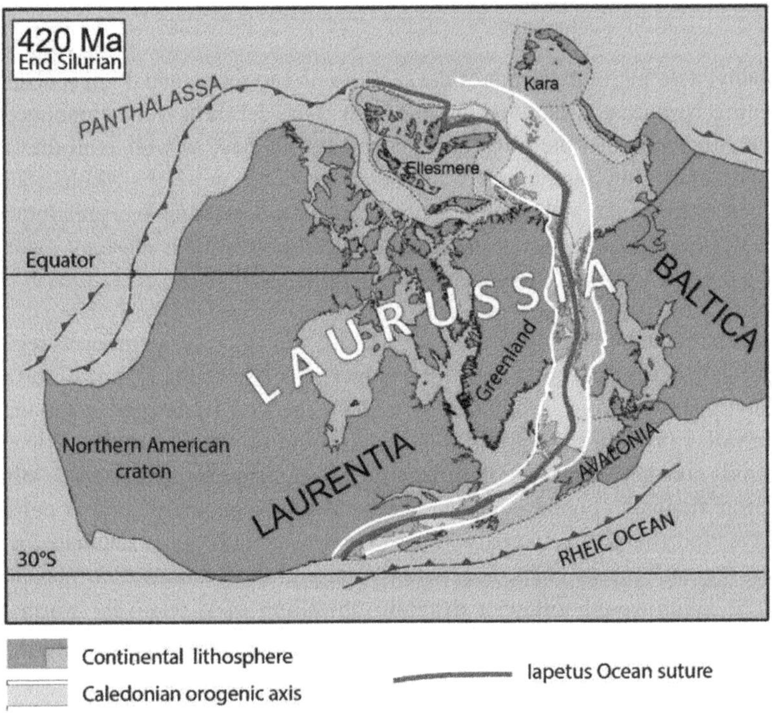

Figure 3.2. *Laurussia at the end of the Silurian (from Cocks and Torsvik 2011)*

3.2.1. *Passive margins and large igneous provinces*

The "non-volcanic passive margins" are those where the continental extension process is not associated with significant mantle melting. This process is generally slow, lasting tens of millions of years. These "cold" margins are characterized by the absence of significant volumes of mafic magmas in the crust during extension, and by the late exhumation of continental sub-lithospheric mantle in the distal zone of the margin (Unternehr *et al.* 2010) (Figure 3.3(a)). In the proximal parts of these margins, the upper (brittle) crust is cut by faults dipping dominantly toward the ocean. These controlled the deposition of sediments, known as syn-rift sediments.

The "volcanic passive margins" (Figure 3.3(b)) formed in a context of very significant mantle melting in opposition to the non-volcanic ones. This process occurred during continental extension from an adiabatic decompression of the mantle (White *et al.* 1987; Geoffroy 2005). The structure of passive volcanic margins is very different from that of non-volcanic margins. The extension occurs in the crust by tectonics, but also by the continuous addition of magma in the form of dykes, which substantially modifies the rheology of the crust and the lithosphere during its stretching and thinning (Annen and Sparks 2002; Callot and Geoffroy 2004).

These margins are thus not only continent-ocean transitions associated with extensional deformation but also sites of magmatic accretion in a continental context. This accretion results in an increase in volume of the extending crust. Thus, the crustal thickness of these volcanic margins is significantly greater than that of non-volcanic margins (Figure 3.3(b)), although the tectonic elongation and thinning of the initial continental crust are significant. Nowadays, an excellent documentation of the structure of volcanic passive margins exists at crustal scale, resulting from advances in wide-angle seismic reflection. This technique allows imaging the structure of margins down to Moho and beyond. These data clearly show that, in contrast to non-volcanic passive margins, no translithospheric shear with mantle exhumation can be deduced from the study of volcanic passive margins.

Deformation in the crust of passive volcanic margins is accompanied by the development of major tectonic flexures with continuous filling of lavas emplaced under subaerial conditions and dipping toward the oceanic crust. These lavas form, in seismic reflection, prisms (or wedges) of seaward-dipping reflectors (SDRs) (Figures 3.3(b), 3.4, and 3.12(f) and (g)). The SDRs lava prisms may be of considerable thickness, (sometimes up to 10–12 km) and were created in a short time, probably less than 4 My (Lenoir *et al.* 2002) (Figure 3.4).

A distinction is made by their geometry between SDRs located in the internal position of the margin (i.e. toward the continent), and those located in a distal position (close to the continent-ocean boundary). Planke *et al.* (2000) called them inner and outer SDRs, respectively. The setting of inner SDRs occurs clearly in the continental domain and is subaerial. The interpretation of outer SDRs is more controversial, although Planke *et al.* (2000) also attribute them to the continental margin. The lava series in outer SDRs have a lesser thickness, and the tip of the reflectors corresponds to a sub-horizontal surface, at the roof of a crust whose continental or oceanic nature is not known with certainty (Planke *et al.* 2000; Franke *et al.* 2013; Geoffroy *et al.*, 2020) (Figure 3.3(b)).

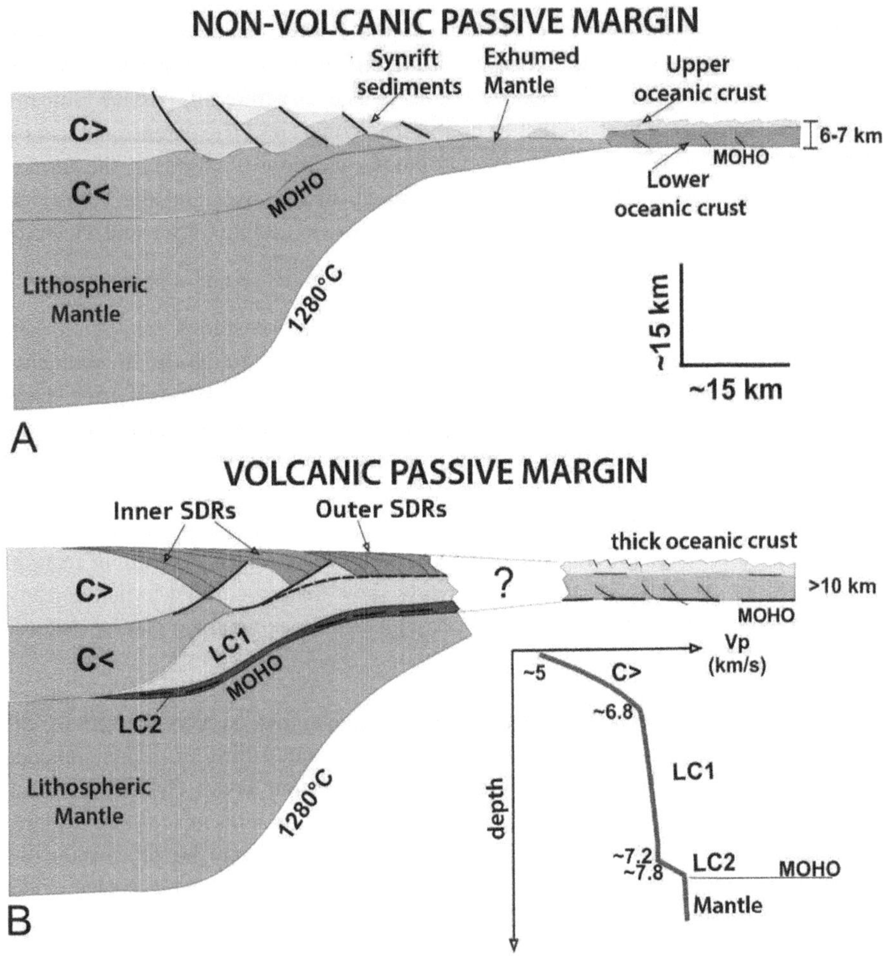

Figure 3.3. *The two major modes of continental breakup*

COMMENT ON FIGURE 3.3.– *(A) Passive non-volcanic margins. (B) Volcanic passive margins. Not to scale. No post-rift sediments indicated. C > and C <: upper and lower* *crust, respectively. LC1 and LC2 lower crusts with high P-waves velocities, injected with magma in the form of intrusions (sills). The diagram shows schematically the evolution with depth of the seismic P-wave velocities (shown on the curve) from the SDRs to the MOHO through a volcanic passive margin. Note the particularly weak velocity gradient at the level of the lower crust LC1 which is, moreover, often very thick. The thicknesses are relative in the diagram.*

Observations of very deep seismic profiles suggest that this crust could be a continental lower crust, highly injected with magma, and exhumed during the extension process (Geoffroy *et al.* 2015, 2020). We will discuss this point later, as it is of direct interest to the tectonic and geodynamic interpretation of Iceland and its immediate surroundings (section 3.3). It should also be noted that the upper crust of the outer part of the volcanic passive margins can also contain lava flows in a sub-horizontal position (FLF [flat-lying flows]-type crust in this book) (Soto *et al.* 2011). FLF crust is seismically distinct from the oceanic crust by its smooth upper surface and its larger thickness. Its interpretation remains enigmatic, although it is possibly similar to the crust with outer SDRs (Geoffroy *et al.* 2020).

The inner SDR prisms of the volcanic passive margins are controlled by listric detachments dipping toward the continent, a pattern totally different from that observed at the non-volcanic passive margins (Figures 3.3 and 3.4). SDRs thus form syn-tectonic lava piles filling the top of anticline roll-overs (Geoffroy 2005; Clerc *et al.* 2015). In the deeper crust, deformation is under control of ductile flow and of the injection of large volumes of magma, essentially in the form of planar intrusions (dykes and sills) (Figures 3.3(b) and 3.4). The magma in the upper crust originates essentially from magmatic chambers of large polygenic volcanoes (shield and composite in type), themselves fed by mantle melting (Callot and Geoffroy 2004; Geoffroy 2005).

At the scale of a single volcanic passive margin, rates of mafic magma emplacement within the crust (lava flows and intrusions) can exceed 1 km^3/year (Eldholm and Grue 1994). These margins generally develop within large igneous provinces (LIPs). LIPs are associated with the extrusion and intrusion of millions of km^3 of magma in a generally short time, sometimes only a few million years (Courtillot *et al.* 1999).

Mantle melting may precede the formation of these margins – and thus the continental extension – and affect a large area with irregular contours (Coffin and Eldholm 1994). This early volcanism constitutes the traps (or "plateau basalts") that can cover very large areas, greater than 10^6 km^2. Significantly, these traps are subhorizontal and can be located on normal-thickness continental crust, demonstrating the absence of extensional (or compressive) tectonics during their emplacement (Geoffroy 2005).

LIP magmatism is often interpreted as the consequence of adiabatic decompression of a warmer and/or more fertile than the normal sub-lithospheric mantle, able to thin thermally and melt the base of the lithosphere. The mafic magmas of the LIPs have chemical compositions enriched in incompatible elements (large cations and/or high charge cations, difficult to incorporate in silicates) compared to the depleted MORB (Mid-Ocean Ridge Basalts)-type magmas produced at the level of oceanic ridges.

This observation led to a mantle model consisting of two large reservoirs: (i) the upper mantle, progressively chemically depleted since the Archean by extraction of incompatible elements (formation of the continental crust), and (ii) the "primitive" lower mantle, not depleted. This model considers LIPs as the expression of deep plumes of mantle origin, born of instabilities at the interface between the earth's core (D'' layer) and the lower mantle (Campbell 2005).

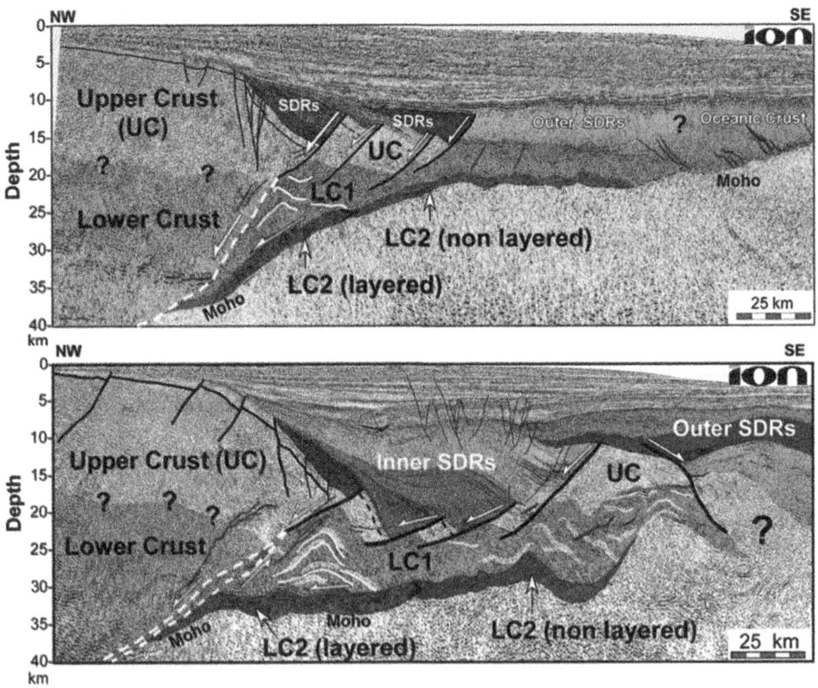

Figure 3.4. *Interpretation of seismic reflection profiles off the coast of Brazil (Pelotas volcanic margin) (from Geoffroy* et al. *2015)*

COMMENT ON FIGURE 3.4.– *Compare with Figure 3.3(b). Note the significance of SDRs in the upper crust. LC1 and LC2: different facies of lower crust more or less injected with magma . UC: upper crust. Although data of this quality do not yet exist in the Northeast Atlantic in seismic reflection, the nature and structure of the passive margins in this region is close to those described in this figure.*

We need to distinguish between the terms mantle plume and hot spot, which are often confused in the scientific literature, a feature which can lead to misunderstandings (Courtillot *et al.* 1989; Richards *et al.* 1989). Plumes theoretically consist of large-diameter, regional sub-lithospheric thermal anomalies related to the gravitational instability of ascending deep warm mantle and to its spreading beneath the lithosphere (plume head) (Campbell 2005) (Figure 3.5). Large trap-covered areas are classically associated with adiabatic partial melting of the plume head, which thermally erodes the base of the lithosphere. Long wavelength lithospheric uplift is considered to be related to the ascending mantle (active) flow and the increased isostatic buoyancy of the lithosphere, resulting from thermal thinning and high heat fluxes (Saunders *et al.* 2007). Plume heads may cool if the upward mantle flow decreases. Only the rising axis (plume tail) may then remain temporarily active and give rise to a more local thermal anomaly, a hot spot (Richards *et al.* 1989) (Figure 3.5). But this relationship of hot spots with large-diameter plumes is only one possible interpretation of localized zones of mantle melting under the lithosphere. Courtillot *et al.* (2003) suggest that only a small number of active terrestrial hot spots (seven out of 49, including Iceland) have the characteristics of a very deep origin (layer D'').

The simple two-layered Earth mantle resulting model, as well as that of deep mantle plumes, has been challenged by seismic mantle tomography imagery that shows the importance and complexity of the material exchange between the upper and lower mantle (Van der Hilst and De Hoop 2005). In addition to subductions of oceanic lithosphere slabs (Li *et al.* 2008), many studies point to the importance of small-scale convective processes in upper mantle dynamics (Davaille and Jaupart 1993; Dumoulin *et al.* 2005).

We will resume this discussion on plumes with the aim of explaining the processes occurring from the early opening of the North-East Atlantic to the formation of present-day Iceland.

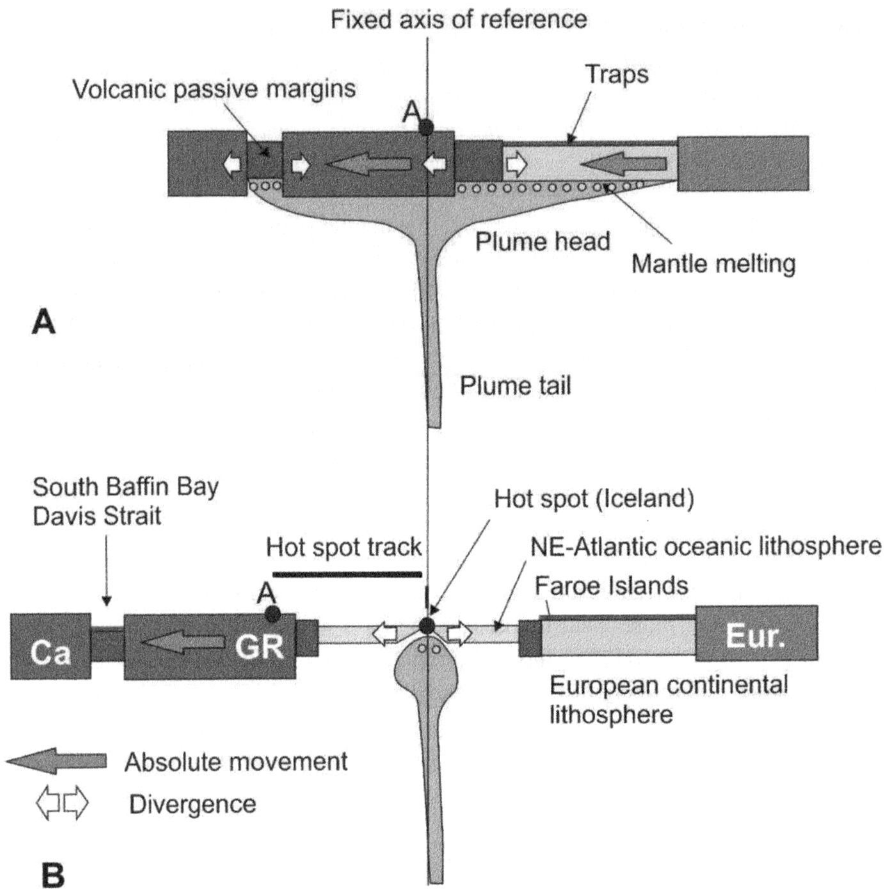

Figure 3.5. *Model of a mantle plume rising beneath a lithosphere of variable thickness and LIP with application to the Northeast Atlantic region*

COMMENT ON FIGURE 3.5.– *(A) Spread of the plume head under the lithosphere with partial melting by adiabatic decompression (yellow dots) at thinner lithosphere zones (Thompson and Gibson 1991), formation of traps followed by volcanic passive margins in diverging zones. (B) Cooling of the plume head with temporary maintenance of its rising axis forming a hot spot . From A to B, the plates and the ridge (Figure 2.1) drift absolutely westward above the rising axis of the plume (tail), which is considered fixed. The application of this model to the North Atlantic region will be discussed in section 3.2.5.*

3.2.2. *The early beginnings of the opening of the North Atlantic Ocean*

3.2.2.1. *Magnetic anomalies, magnetochrons and types of lithosphere*

The continental lithosphere is distinguished from the oceanic lithosphere by the nature of its crust, continental or oceanic. Because of its mafic composition, rich in ferromagnesian minerals, and of the symmetrical accretion of the oceanic lithosphere on either side of the oceanic ridges, the oceanic crust is marked by pairs of linear magnetic anomalies of increasing age as one moves away from the accretion axes. This feature distinguishes it from continental crust.

As the polarity of the Earth's magnetic field changes (irregularly) over geological time, these anomalies form symmetrical bands, of normal or reverse polarity, on either side of the accretion axes (mid-oceanic ridges). These anomalies, combined with absolute dating of the rocks, provide a magneto-stratigraphic scale (Figure 3.6). They then correspond to durations of geological time, thus defining "magnetic chrons" of symbol C and numbered from the most recent (C1: Current) (Ogg *et al.* 2016). Each chron generally corresponds to a pair of magnetic anomalies, inverse then normal (e.g. C13r, then C13n). Moreover, several brief inversions of the magnetic field may be recorded within a chron (Figure 3.6).

The presence of linear magnetic anomalies with alternating polarities is thus very generally considered as a diagnosis of the presence of oceanic crust – and therefore of the corresponding lithosphere – in a submerged domain. This point is now being discussed, as SDR prisms of passive volcanic margins (Figures 3.3(b) and 3.4) can also generate pairs of synchronous anomalies in the stretched continental domain (Stica *et al.* 2013; Koopmann *et al.* 2014; Franke *et al.* 2019; Geoffroy *et al.* 2020).

This point also has a real importance in the geodynamic interpretation of the Northeast Atlantic Ocean (section 3.3).

The key periods for the kinematic and geodynamic history of the Northeast Atlantic Ocean are C33 (Maastrichtian; 70 My), C27 (Middle Paleocene; 61.9 My), C24 (Early Eocene ; 55 My), C18 (Early Bartonian; 40 My), C13 (Early Oligocene; 33.9 My), C6 (Burdigalian-Serravallian boundary; 15 My) and C3 (Early Messinian; 7 My; Figure 3.6). These different periods also correspond in Europe and North Africa to changes in plate kinematics and/or episodes of volcanism or to the opening of straits, exercising major control for thermohaline circulation.

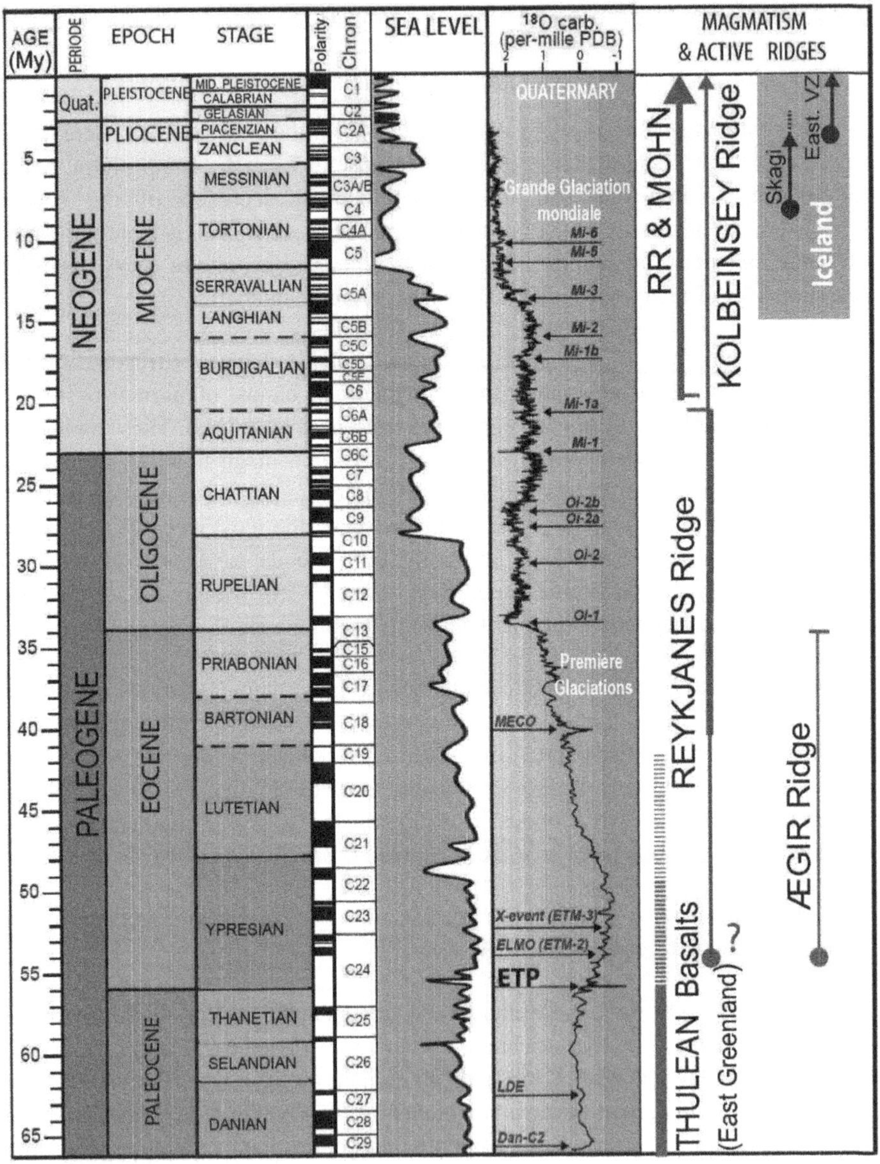

Figure 3.6. *Cenozoic magneto-stratigraphic and climatic scale (Ogg* et al. *2016) and evolution of the Northeast Atlantic (Ellis and Stocker 2014, completed)*

We will describe in the following sections of this chapter the processes at the origin of the progressive breakup of Laurussia after the Caledonian orogeny.

3.2.2.2. *Evolution in West Greenland*

The history of the Northeast Atlantic and Iceland cannot be described without mentioning the geodynamic events that occurred west of Greenland. It is indeed in this zone that the continental breakup and generation of an oceanic lithosphere between the North American and Eurasian plates were initiated during the Upper Cretaceous through the ancient continent of Laurentia (Figure 3.2).

West of Greenland, the Labrador Sea is an area where the axis of ocean expansion is no longer active (Figures 3.7 and 3.8(a)). The extinct ridge is clearly visible in the gravimetric map, as shown in Figure 3.7. The history of continental extension and then breakup between Greenland and Canada in the Labrador Sea remains poorly constrained.

The boundary between Greenland and Labrador (present shapes) has been under extension since at least the Permian period. Alkaline intrusive magmatism, associated with limited volumes of magma, is expressed between 220 and 150 My in southwest Greenland (Larsen *et al.* 2009). At the Lower Jurassic-Cretaceous boundary and up to the Hauterivian (about 150–133 Ma), a swarm of mafic dykes was injected at the future passive margin of Southwest Greenland (Watt 1969; Larsen *et al.* 2009). These dykes are responsible for a low crustal dilatation. However, this demonstrates that an extensive stress regime with minimal horizontal stress orthogonal to the future passive margins was already active at this time.

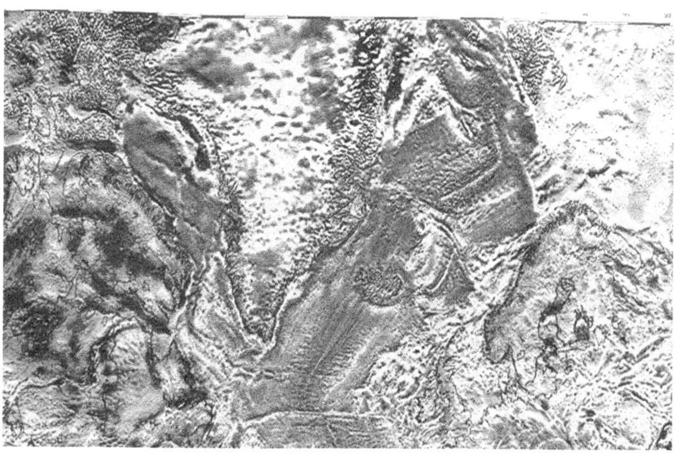

Figure 3.7. *Free-air gravimetric map of the Northeast Atlantic region. Document from the world gravimetric map (Bonvalot* et al. *2012). This map shows the main structural elements of the Northeast Atlantic region, such as ridges, active or extinct, as well as major transform faults*

A first period of Aptian-Albian continental extension is expressed with crustal thinning and stretching along the Labrador–Baffin axis (Oakey and Chalmers 2012). A thermal subsidence followed. These sedimentary basins, Cretaceous in age, are found all along the Labrador–Baffin axis (shown in green in Figure 3.8(a)). The combined passive margins of Labrador and Southwest Greenland are clearly non-volcanic in the southern part of the basin with, for some authors, exhumation of lithospheric mantle (Figures 3.3(a) and 3.8(c)) (Chian and Louden 1994; Chian *et al.* 1995).

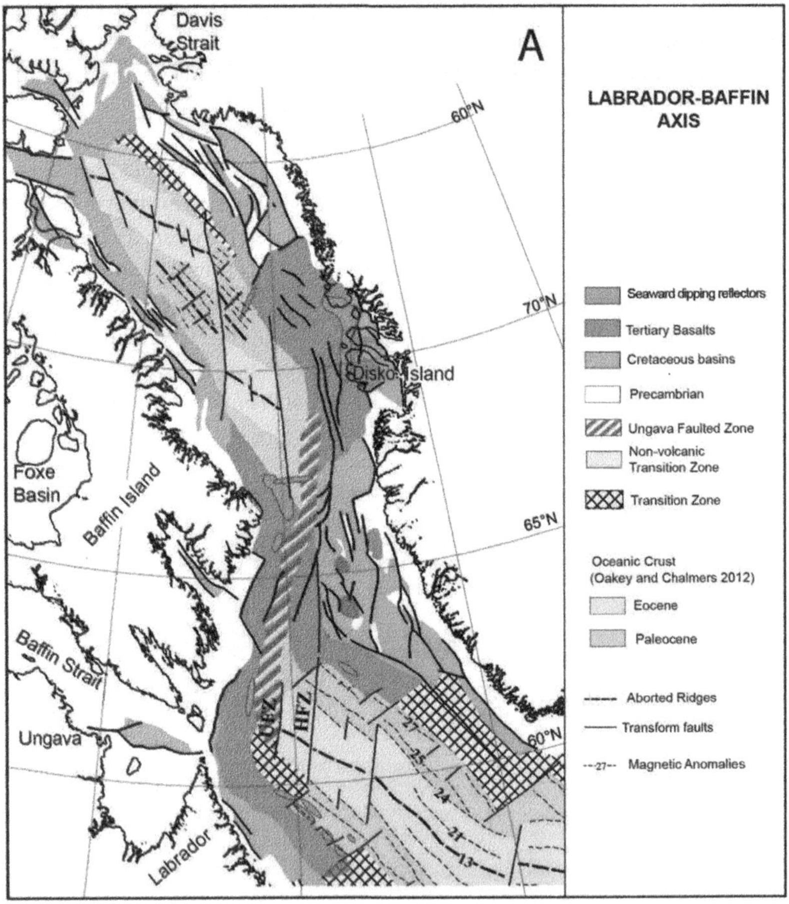

Figure 3.8. *(a) The Labrador–Baffin Rift: geological map of the Labrador Sea and Baffin Bay (Chauvet* et al. *2019 based on Oakey and Chalmers 2012)*

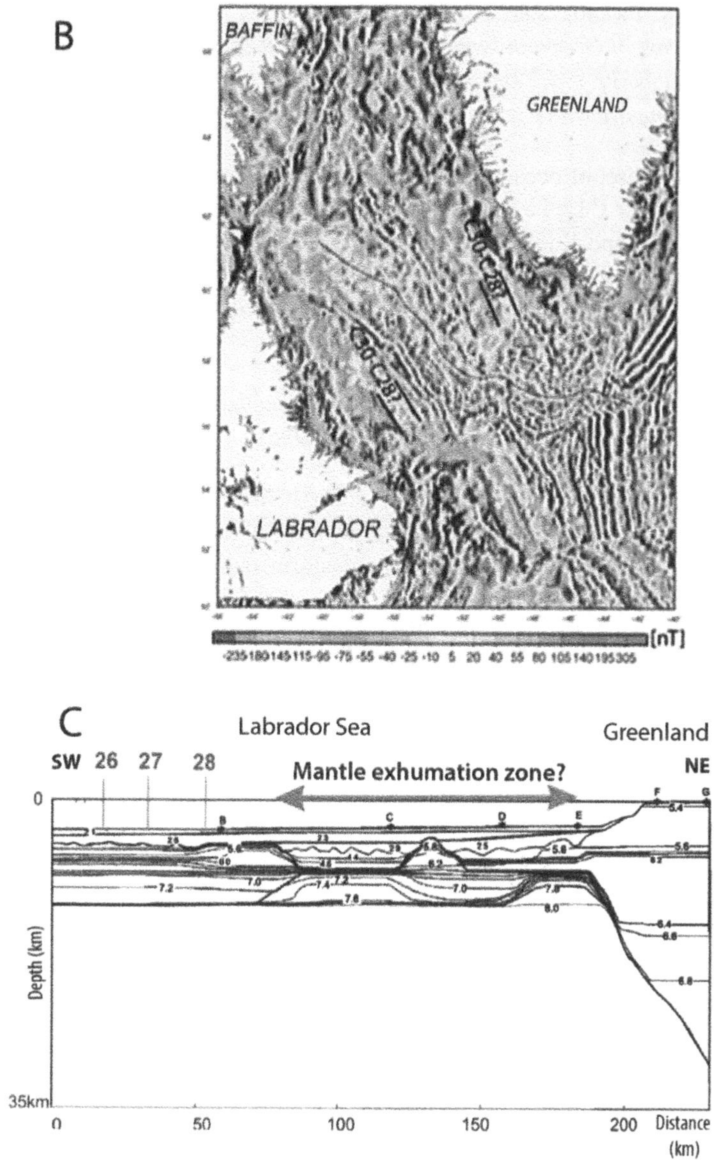

Figure 3.8. *(b) Magnetic anomalies in nannotesla (nT) in the Labrador Sea and Davis Strait (based on Gaina and Jakob 2018). (c) The Labrador-Baffin Rift: southeast refraction seismic profile of the passive non-volcanic Labrador margin with associated magnets anomalies (Chian and Louden 1994)*

COMMENT ON FIGURE 3.8(C).– *Note the transition zone between the thinned continental crust of Greenland and the oceanic crust of the Labrador Sea, with high seismic velocities (Vp > 7.2 km/s), which can be interpreted as exhumed serpentinized peridotites. Compare with Figure 3.3(a).*

The age of onset of oceanic accretion in the southern Labrador Sea is a point of contention: C33 or C31, i.e. late Cretaceous for some authors (Roest and Srivastava 1989; Srivastava and Roest 1999) and C27, i.e. Paleocene (Figure 3.5), for others (Chalmers and Laursen 1995; Oakey and Chalmers 2012). This uncertainty results from the interpretation of magnetic anomalies in the continent-ocean transition zone (Figures 3.8(a)–(c)).

As in the Labrador Basin, oceanic accretion ended in Baffin Bay during the Early Oligocene (C13) (Chalmers and Laursen 1995).

A notable kinematic event in the Labrador–Baffin axis is the change in kinematic direction, which occurred at C25-C24. In the Labrador Sea, it is marked, in particular, by two distinct directions of transform faults (Figure 3.8(a)): they are WSW-ENE oriented between C27 (or C33) and C25 and became SSW-NNE between C24 and C13. This modification follows a relocation of the Eulerian rotational pole between the North American plate and the Eurasian block, and is associated with a temporary acceleration of the divergence between these plates (Roest and Srivastava 1989). It is synchronous with the individualization of a new tectonic plate, the Greenland one. The direction of the Ungava fault system in the Davis Strait is oblique with respect to the two directions of opening identified in the Labrador Sea during the continental extension and following oceanization period C27 to C25, this axis operated in transtension (oblique opening) and then shifted to transpression (oblique compression) during the period C24 to C13. This explains the compressive and extensive structures observed at the level of this transverse axis and which affect, in particular, the Paleogene volcanic formations.

3.2.2.3. *Evolution in East Greenland: a mobile region since the Paleozoic*

Along the Labrador–Baffin axis, Cretaceous extension and continental breakup in the Paleogene affected an ancient Precambrian lithosphere (cold continental lithosphere and therefore, thick and rigid) and propagated, from south to north, obliquely or orthogonally with respect to the ancient Proterozoic sutures (Srivastava and Roest 1999; Schiffer *et al.* 2019).

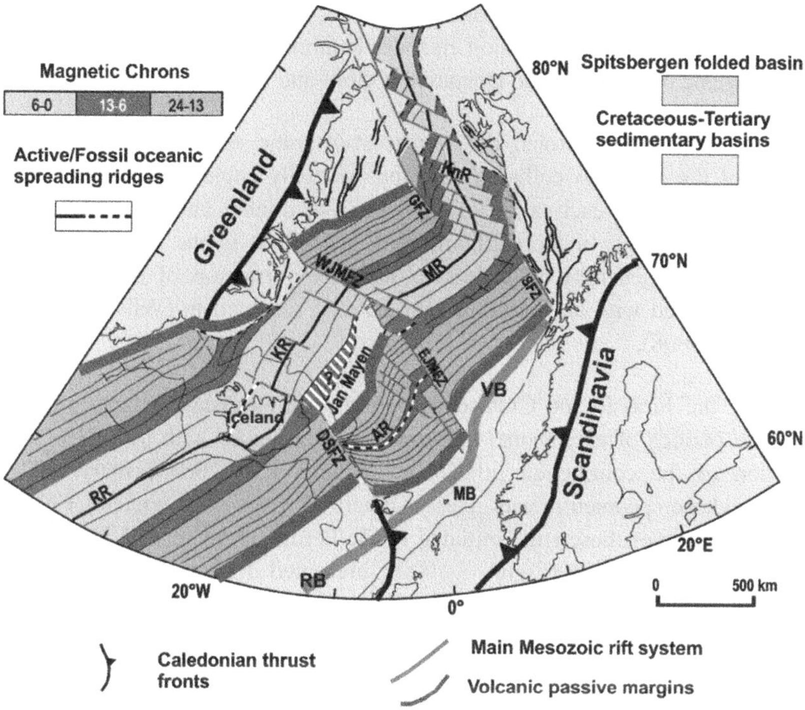

Figure 3.9. *The Northeast Atlantic: main oceanic basins and active or extinct ridges, passive volcanic margins, main Mesozoic rift axis and Caledonian fronts*

COMMENT ON FIGURE 3.9.– *Delineations are approximate and/or simplified (based on Lundin and Doré 2002). RR, KR, AR, MR, KnR: Reykjanes, Kolbeinsey, Aegir, Mohns and Knipovich Ridges. EJMFZ, WJMFZ: eastern and western Jan Mayen fault zones. GFZ: Northeast Greenland fracture zone, SFZ: Senja fracture zone, VB, MB, RB: Vøring, Møre and Rockall basins. DSFZ: Denmark Strait Fault Zone (hypothetical).*

Between Europe and East Greenland, the continental lithosphere did not have the same characteristics: it was a collision chain lithosphere (Caledonian axis, Figures 3.2 and 3.9), much younger (about 400 My) and thus less rigid. Fossil relics of oceanic slabs are discernible under Northeast Greenland and Scotland (Figure 3.8) at the level of the Iapetus paleo-suture (Schiffer *et al.* 2014). This lithosphere, located between the stiffer lithospheres of Greenland and Baltica (Figure 3.2), was easily deformable.

From the end of the Caledonian orogeny, the Caledonian lithosphere was affected by large-scale deformation episodes beginning with a gravitational collapse of the orogenic crust during the Devonian (Gilotti and McClelland 2008).

During the Carboniferous, the Northeast Atlantic region was located in the foreland of the Variscan collision, and thus partially subjected to its deformation field, with a dextral reactivation of the large Caledonian faults, formerly sinistral (Rogers *et al.* 1989). In the Carboniferous as well as in the Permo-Triassic, the lithosphere was generally in extension with the formation of sedimentary rifts, locally associated with an important syn-tectonic magmatism (Wilson *et al.* 2004; Larsen *et al.* 2008).

During the Jurassic and Cretaceous periods, the Northeast Atlantic region was subject to periods of extensional deformation interspersed with periods of thermal relaxation of the continental lithosphere (Figure 3.10). Sedimentary rifts were formed, whose geometry was partly inherited from the ancient Caledonian structures. Some of these rifts, initiated in the Permo-Triassic, such as the Jurassic system of the North Sea (Figure 3.10(a)), presented a three-branch (rift–rift–rift) system with a central dome associated with magmatism (Ziegler 1992). These successive extensive periods profoundly structured the Northeast Atlantic region (Lundin *et al.* 2002). Without exploring the details of these episodes, or the structure of these rifts, sometimes diachronous and of complex geometry (Figure 3.10b), it seems important to note that the maximum extension in this system took place at the end of the Jurassic and at the very beginning of the Cretaceous (Van Wijk and Cloetingh 2002; Osmundssen and Ebbing 2008) (Figure 3.10).

During this period, a quasi-continuous and relatively narrow rift developed from Great Britain to Norway (Rockall Troughs and Møre and Vøring basins), associated with a NW-SE extension and a spectacular lithospheric thinning, locally of the order of 500% (Figures 3.9 and 3.10(a)). The continental lithospheric mantle was probably locally exhumed (Osmundssen and Ebbing 2008). Remarkably, this ancient extension without any significant magmatism did not lead to a continental breakup between Greenland and Europe (Geoffroy 2001; Van Wijk and Cloetingh 2002; Guan *et al.* 2019). Although an extension of the Upper Cretaceous age was suggested locally (Lundin 2002; Gernigon *et al.* 2003), a large part of the Northeast Atlantic region was subject for more than 70 My to a period of generalized thermal relaxation marked by thick post-rift deposits (Figure 3.8(b)).

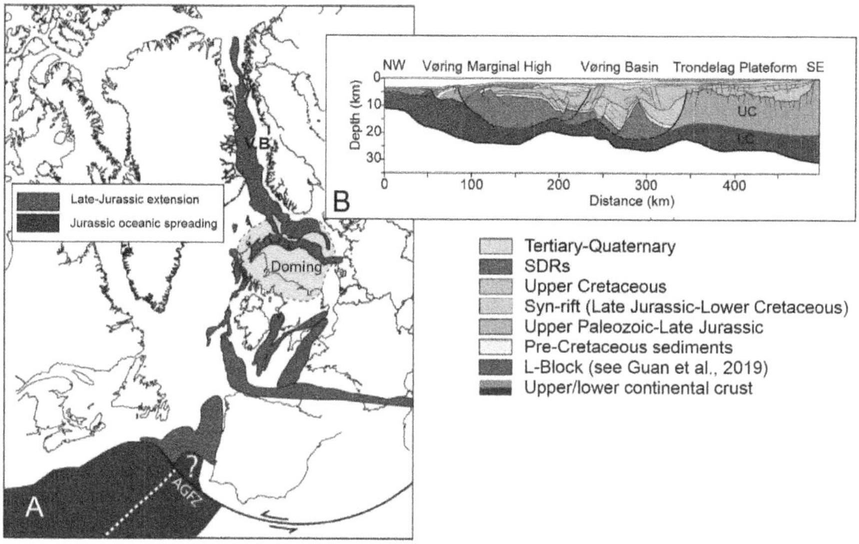

Figure 3.10. *Mesozoic continental extension in the Northeast Atlantic*

COMMENT ON FIGURE 3.10.– *(A) Reconstruction of the N-Atlantic during the Late Jurassic and location of the main rifts and areas with oceanic lithosphere. (B) Mesozoic Vøring Rift off Norway (Guan* et al. *2019, modified). Note the Late Jurassic/Lower Cretaceous age extensional deformation and the very significant crustal thinning induced by this deformation. The post-rift Cretaceous deposits here are affected by normal faults, related to moderate and localized extension, which is not the general case in the Northeast Atlantic. This extension is frequently associated with local domes and has been interpreted by some authors as the consequence of regional compressive or transpressive deformations (Lundin* et al. *2013). The passive volcanic margin developed in the Paleogene west of the Mesozoic rift (Gernigon* et al. *2019; Guan* et al. *2019).*

3.2.3. *Thulean magmatism in the Paleocene and the continental breakup of the Northeast Atlantic*

3.2.3.1. *Thulean traps*

During the Paleocene (C28N to C26R) (Wilkinson *et al.* 2016) (Figure 3.11(b)), the Northeast Atlantic region was affected by a spectacular event, the effusion of mafic magmas in the form of lava flows covering both the eroded bedrocks of Precambrian or Caledonian age and the Mesozoic sedimentary basins (Figure 3.11).

Although now partially eroded, these "plateau basalts" (or traps) can be seen in Northwest Britain (Hebrides, Ulster; Figure 3.12(a-b)) and in the Faroe Islands. The oldest ages are found in the British Isles (Figure 3.11(b)). They are also widely present offshore (Hebrides, Hatton, Rockall Bank, etc.), covering in angular unconformity the Mesozoic sedimentary basins or the Caledonian bedrock.

Volcanism is present on both sides of Davis Strait, west of Greenland (Cape Dyer on Baffin Island and the Disko-Svartenhuk region in west-central Greenland; Figures 3.11(a) and (f), and 3.12(b), (e), and (g)) and east of Greenland (Figures 3.11(a) and (f)). Volcanism in western Greenland seemed, however, to start synchronously with that of the Hebrides, whereas the beginning of volcanism in East Greenland and the Faroes was slightly more recent (about 3 My difference).

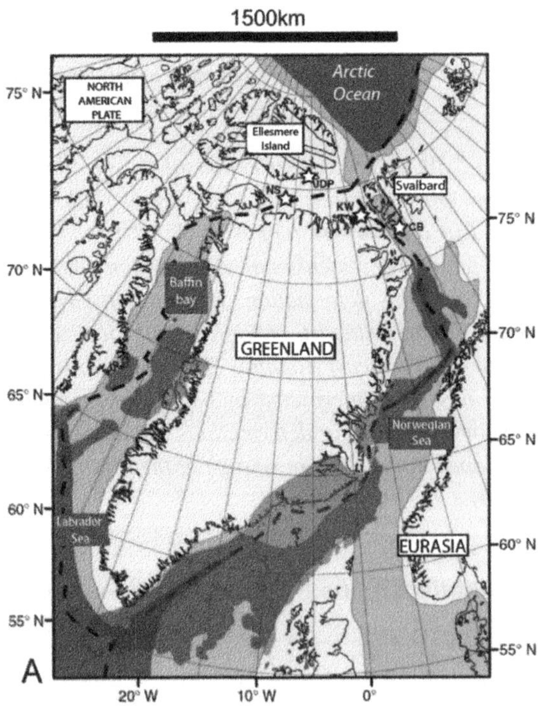

Figure 3.11. *(a) Reconstructed map of the North Atlantic (or Thulean) Paleogene magmatic province. In dark red, magmatic centers (Jones* et al. *2016)*

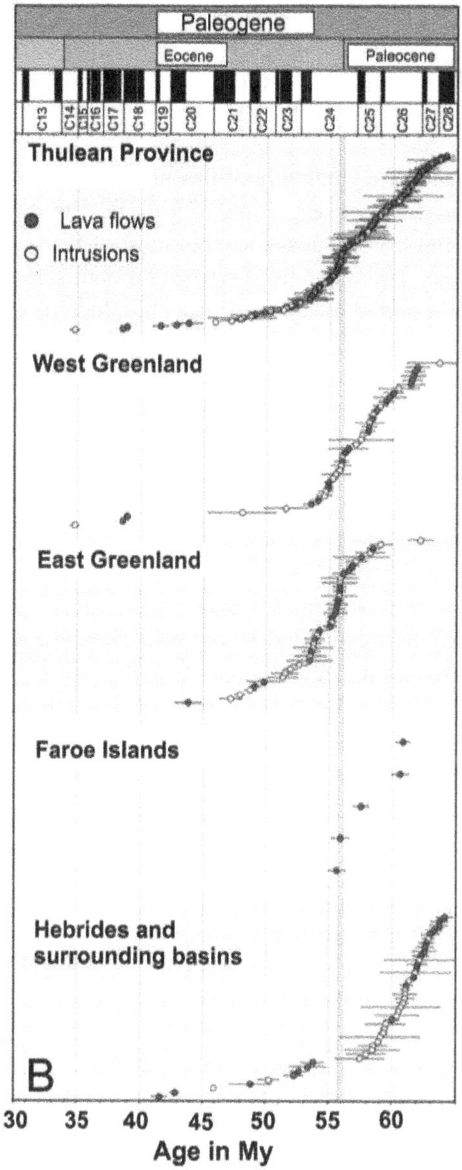

Figure 3.11. *(b)Thulean magmatism: cumulative frequency diagram of North Atlantic basalt dating (from Wilkinson* et al. *2016)*

COMMENT ON FIGURE 3.11a.– *The lava flows on this map (in light red and orange) were emplaced in subaerial context and correspond to either traps or SDRs. Note also the existence of numerous magmatic centers (in dark red). See also Ziegler (1989) and Scotese and Scotese (2006).*

There is no argument for a continuity of this magmatism between West and East Greenland under the present ice sheet. Some of these basalts may be alkaline (e.g. in Scotland, at the beginning of the emissions), but most of the volume is composed of tholeiitic basalts (Chapter 1 of Volume 2) emitted on both sides of Greenland. Their emplacement followed a period of uplift of the lithosphere during the Maastrichtian and Danian periods. This uplift led to syn-sedimentary incisions in existing basins on either side of Greenland (Dam *et al.* 1998) as well as to influxes of detrital material into residual basins around the Hebrides (White and Lovell 1997). The basalts that can be observed in the Northeast Atlantic region were emitted from the beginning of the regional magma event either subaerially (Figure 3.12 (a), (b), (f) and (g) or in shallow settings (hyaloclastite lake filling: Figure 3.12(e)), a feature consistent with this pre-magmatic regional uplift.

Almost all Thulean volcanism is of fissural origin, the lavas being emitted from vertical dykes, generally between 1 and 10 m thick. The dyke swarms that fed these lava emissions originated from magmatic centers of the same age and large volumes, located within the upper crust, at paleo-depths of 2–3 km (Callot and Geoffroy 2004). These magmatic centers (Figures 3.11(a) and 3.17), located under large polygenic (composite) volcanoes, collected the magma resulting from mantle melting.

Most of the dykes observed in the magmatic province correspond to a hydraulic-type rupturing around these intracrustal reservoirs, with magma injecting itself laterally as dykes, sometimes over tens of kilometers. The intersection of these dykes with the topographic surface was thus able to supply lava to the magmatic province at some distance from the magmatic centers (Geoffroy *et al.* 2007).

Onshore observations in West Greenland, Ulster and the Scottish Islands of the Hebrides show that the continental extension, during the onset of Paleocene volcanic trap emplacement in the North Atlantic region, was very small, outside the continental breakup axes located on either side of Greenland (section 3.2.3.2). Based on the dilation associated with the intrusion of the dykes, this Paleocene age extension is estimated at only 1% in the Hebrides, except in the immediate periphery of the magmatic centers where it can reach 10–15% during tholeiitic volcanism (Mattey *et al.* 1977).

Figure 3.12. *(a and b) Paleocene-Eocene magmatism of the Northeast Atlantic region. (A) Traps with paleosols (Ulster, near Giant's Causeway). (B) Picritic traps of Paleocene age (Greenland, Nuussuaq Peninsula) (© Laurent Geoffroy)*

Figure 3.12. *(c and d) Paleocene-Eocene magmatism of the Northeast Atlantic region. (C) Early Eocene dyke swarm in Southeast Greenland cutting through the Greenlandic Precambrian bedrock after erosion of the inner SDR located above this injected crust. Dyke dilation of the crust can reach 50%. (D) Paleogene sill in Cretaceous sediments of Northeast Greenland (© Laurent Geoffroy)*

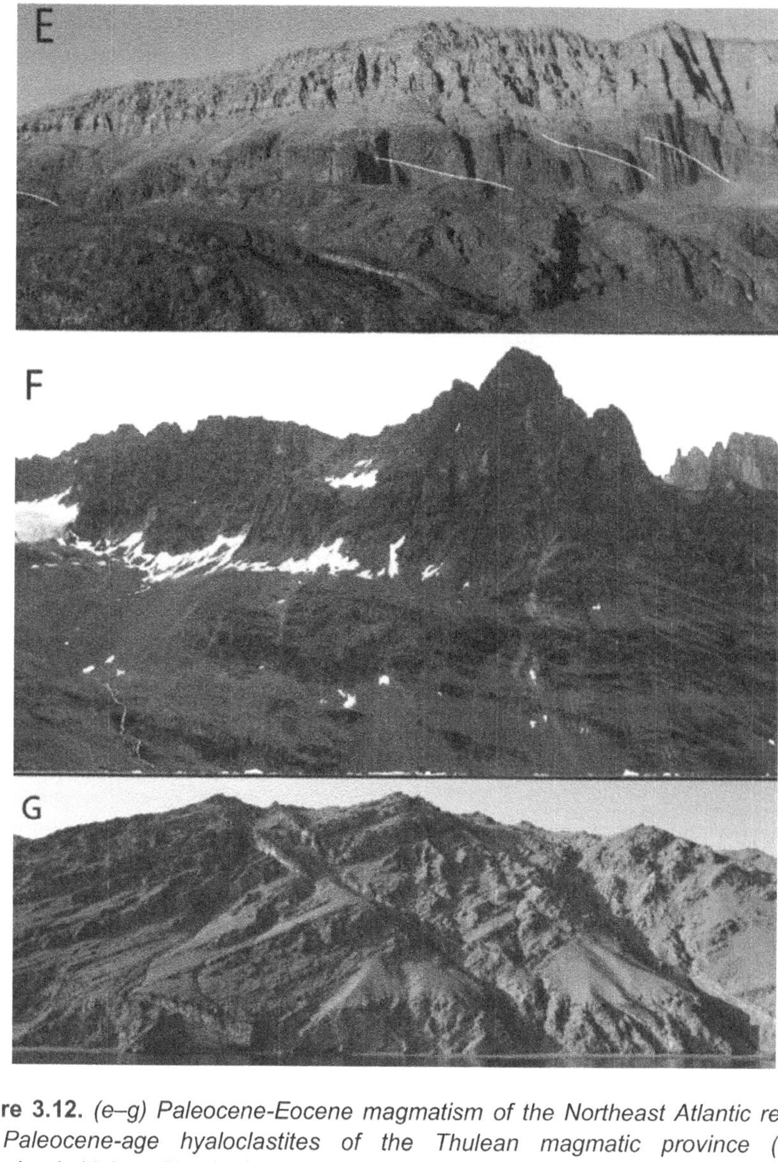

Figure 3.12. *(e–g) Paleocene-Eocene magmatism of the Northeast Atlantic region. (E) Paleocene-age hyaloclastites of the Thulean magmatic province (West Greenland, Vaigat Fjord). (F) Inner SDR of the Blosseville Coast (East Central Greenland). The lava flows are syn-tectonic and dated at about 56 My. (G) Inner SDR (see Figure 3.3b) of Paleocene age in West Greenland (Svartenhuk) (© Laurent Geoffroy)*

COMMENT ON FIGURE 3.12(e-g).– *(E) Hyaloclastites (picrites) (section 1.3.3 of Volume 2) fill up a residual Danian-age lake and are covered by aerial lava flows (also picritic, see Figure 3.12 (a) and (b)). These formations, which are discordant over the Cretaceous sedimentary basin, underwent no deformation during the lava emplacement (except for the dilatation associated with the injection of dykes into the crust) (© Laurent Geoffroy). (F and G) Fan-shaped Inner SDRs are associated with significant stretching and thinning of the continental crust.*

3.2.3.2. *Paleogene syn-magmatic breakup*

The continental breakup in the whole Thulean province was thus realized in the context of mantle melting during the Paleogene. The passive margins were all of volcanic type in the Northeast Atlantic, as in southern Baffin Bay (Figures 3.9 and 3.13). They were associated with the development in the upper crust of SDR prisms (Figure 3.12 (f), (g), section 3.2.1) whose thickness and distribution are not known with certainty. West of Greenland, the continental breakup was also magmatic in the Davis Strait (transtensive period) and in the southern Baffin Bay. Continental extension began in west Greenland during the Early Paleocene (C27) after a brief episode of subaqueous hyaloclastite (Figure 3.12(e)) and trap (Figures 3.12(a), (b)) emplacement, and continued during the Eocene (Chauvet *et al.* 2019). The precise age of the continental breakup and the beginning of oceanization in the Baffin Bay is not known with certainty, but it is certainly synchronous with or later than C24. East of Greenland, the extension and then the continental breakup were also of Eocene age.

Everywhere, syn-magmatic breakup was associated with magmatic volumes and mantle melting rates greater than at the initial stage of traps setting, of the order of 1 km^3/year over a short deformation period, probably less than 7 My (Eldholm and Grue 1994; Wilkinson *et al.* 2016). The rates of mantle melting increased with lithospheric thinning by extension (Agranier *et al.* 2019) and reached values close to those of oceanic ridges. The magmatism of these margins was not only associated with the extrusion of SDRs (Figures 3.12(f) and (g)), but also with the intrusion of similar volumes of magma into the stretched continental crust located below the SDRs. These intrusions were dykes within the upper crust (Figures 3.12(c)) (Klausen and Larsen 2002; Lenoir *et al.* 2003) and sills within the lower continental crust (White *et al.* 2012) corresponding to LC1 and LC2 in wide-angle seismic (Figure 3.3(b)).

The sills were also injected into the entire Cretaceous post-rift sedimentary sequences of the Northeast Atlantic region (Figure 3.12(d)). They caused a significant thermal outgassing of CO_2 contributing to the warm PETM event (thermal maximum at the Paleocene/Eocene boundary; section 2.1 of Volume 2).

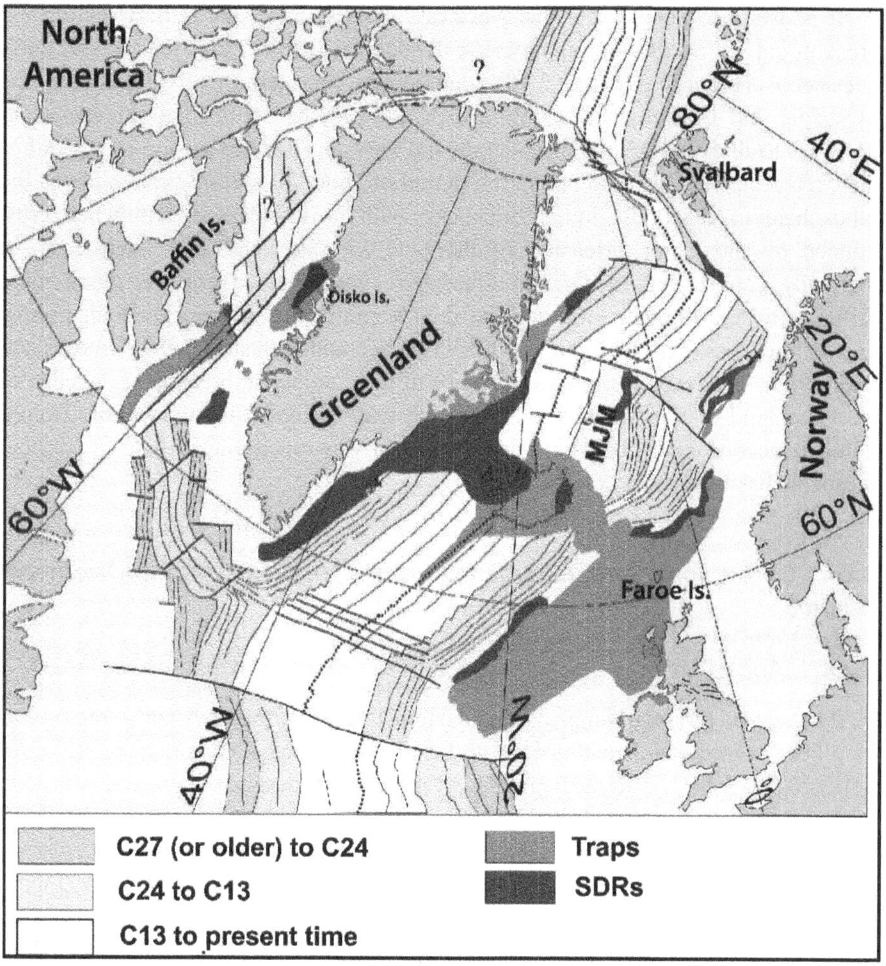

Figure 3.13. *Volcanic margins of the Northeast Atlantic (from Chauvet* et al. *2019). MJM: Jan Mayen microcontinent*

The passive margins of the Northeast Atlantic domain were associated with a relatively thick crust. Indeed, the crust increased in volume by the addition of magmatic intrusions and extrusions. However, it was stretched and thinned in relation to tectonic extension, and its final thickness at the time of continental break-up remains greater than that of a passive margin without the addition of magma during deformation (Figure 3.4).

It is useful to note that, in the Northeast Atlantic axis, the Tertiary continental breakup did not occur along the hyper-extended rifts of the Late Jurassic and Early Cretaceous (Figures 3.9 and 3.10), due to the long-term interval between the Mesozoic and Cenozoic extension, which allowed the continental lithosphere to cool, i.e. to thicken. Despite a moderate and very localized extension at the end of the Cretaceous, this thermal relaxation period of about 70 million years allowed the lithosphere to regain its initial thickness, with nevertheless a continental crust thinned by the major extension of the Late Jurassic and Early Cretaceous. A thermally equilibrated continental lithosphere with thinned crust is much more difficult to deform again in extension than a continental lithosphere with normal crustal thickness, because it is less ductile (Brace and Kolhstedt 1980; Kusznir and Park 1989). This explains the Eocene shift of the continental breakup at the edge of the Greenland craton, where the lithosphere was not only slightly warmer (minor Late-Cretaceous extension), but where the crust was much thicker than in adjacent Mesozoic basins (Guan *et al.* 2019).

3.2.4. *Chronology and kinematics of the opening of the Northeast Atlantic*

3.2.4.1. *Age of onset of oceanization in the northeast Atlantic*

The age of the beginning of oceanization in the Northeast Atlantic region has long been considered as dating back to C24 based on magnetic anomalies (Figures 3.9, 3.13 and 3.14), i.e. to the beginning of the Eocene, despite the presence of SDRs in the crust, dated at C24. Initial concepts on passive volcanic margins considered SDRs as related to a particular oceanic accretionary process based in particular on the observation of SDRs in Iceland (section 3.3.2). However, the inner SDRs (Figure 3.3(b)) exposed ashore in Greenland in the most proximal part of the volcanic margins (Figure 3.12(g)) are clearly in continental position. At sea, the inner SDRs related to the first oceanic magnetic anomalies were drilled during ODP campaigns. The geochemistry of the lava forming these SDRs clearly demonstrates their contamination by continental material (Larsen *et al.* 1998).

As discussed in section 3.2.2.1, both inner and outer SDRs give rise to linear magnetic anomalies, often with higher amplitudes than oceanic anomalies (Franke *et al.* 2019). It is therefore difficult, in the context of magma-induced continental breakup, to use magnetic anomalies to locate the continent-ocean boundary. We will come back to this limitation in section 3.3, as it has major consequences for the interpretation of Iceland.

3.2.4.2. *The ocean basins of the Northeast Atlantic*

Figure 3.14 shows the distribution of magnetic anomalies in the Northeast Atlantic. Linear anomalies are well expressed in most of the basins or sub-basins that constitute it. They can be used as benchmarks to reconstruct its kinematic evolution (Figures 3.9 and 3.15), with the limitations expressed above.

East of Greenland, magnetic anomalies, like gravimetric anomalies (Figure 3.7) delineate distinct ocean basins within which oceanic accretionary processes are still active, with the exception of the Ægir fossil basin (or ridge) in the Norwegian Sea (Figure 3.9). The oceanic crust of the Northeast Atlantic Ocean is never described as older than C24 (corresponding to the Early Eocene, Figure 3.6), from the large Charlie–Gibbs transform fault, west of Ireland, to the Arctic Ocean (Figures 1.5 and 3.5). This illustrates the fact that the first continental breakup axis of the North Atlantic was located west of Greenland, within the Labrador–Baffin axis (Figure 3.9), although the northeast Atlantic domain east of Greenland has been subject to successive periods of continental extension since the Paleozoic (section 3.2.2.3).

The interdependence of the two opening systems west and east of Greenland is in fact obvious: the beginning of oceanic expansion in the Northeast Atlantic system was contemporary with – or consecutive to – the kinematic reorganization in the Labrador Sea at the Paleocene-Eocene boundary (C25-C24; section 3.2.2 and Figure 3.8(a)). This was the pivotal period of the individualization of a new lithospheric plate, Greenland, which was bounded to the northwest (Nares Strait) as well as to the northeast (Fram Strait) by transform fault systems with transpressive sinistral and dextral shear, respectively (see, for example, the Spitsbergen folded basin in Figure 3.9).

This plate drifted northward during the C24-C13 period (which covered approximately the duration of the Eocene, 22–24 My) due to simultaneous oceanic accretion at the two accretionary axes Labrador–Baffin and Northeast Atlantic.

This northward drift also led to an intracontinental compressive deformation in the Canadian High Arctic, corresponding to the Eureka chain (61–45 My), expressed notably on Ellesmere Island (De Paor *et al.* 1989; GCMCC 2018 Tectonic Map of the Arctic). Greenland's drift ceased at the end of oceanic accretion in the Labrador–Baffin axis at C13 (beginning of the Oligocene, at 33–34 My). From C13, Greenland and the Labrador–Baffin axis became part of the North American plate.

We describe below the evolution of the divergence between Greenland and Europe during the Cenozoic based on the study of the different ocean basins of the Northeast Atlantic.

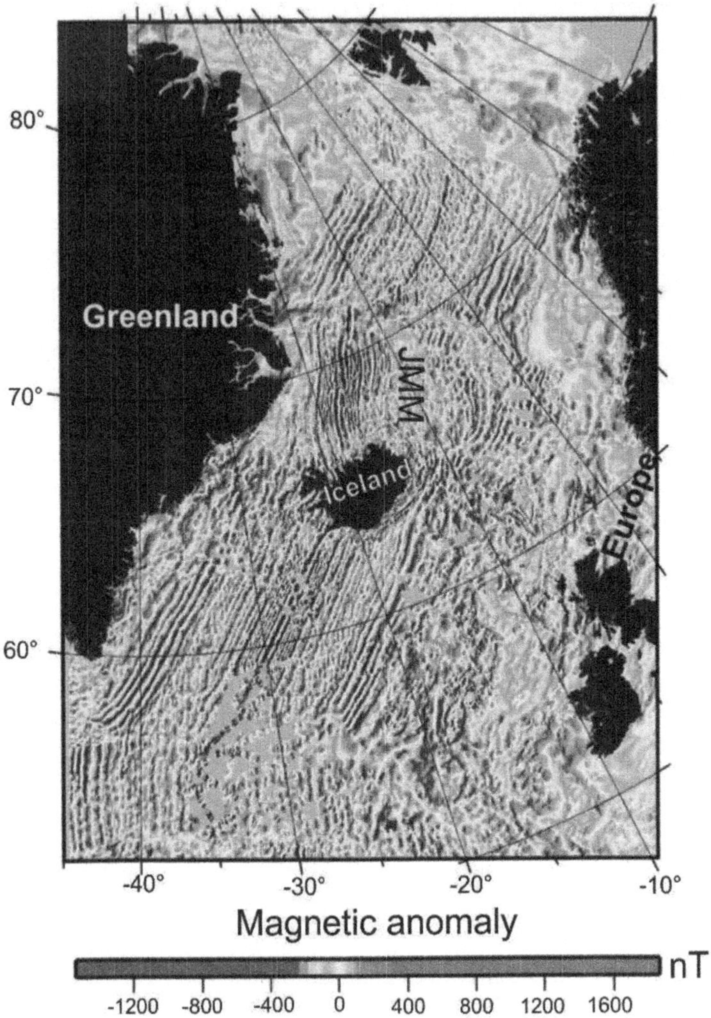

Figure 3.14. *Magnetic anomalies in the Northeast Atlantic Ocean (Gaina* et al. *2017). JMM: Jan Mayen's microcontinent*

3.2.4.3. *The Reykjanes Basin in Southern Iceland*

Magnetic anomalies at the Reykjanes Ridge (Figures 3.9, 3.13, 3.14 and 3.15) suggest a continuous accretion from the ridge from C24 to the present.

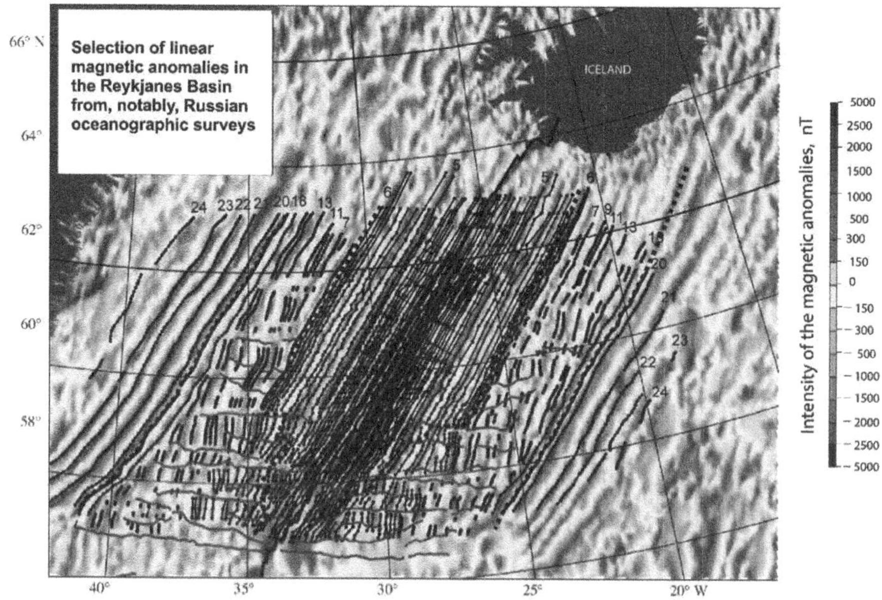

Figure 3.15. *Magnetic anomalies in the Reykjanes Basin (Mercur'ev* et al. *2009)*

In detail, however, these anomalies are of distinct expression according to their age (Figure 3.15): from C24 to C18, they were continuous and parallel to the volcanic margins of southeast Greenland and Hatton Bank, and associated with a NW-SE opening.

This arrangement changed from C18, at the beginning of the Bartonian (41 My): the anomalies became strongly segmented at a small scale and of N-S direction at segment scale. They were orthogonal to an E-W direction opening (Mercur'ev *et al.* 2009; Martinez *et al.* 2019). This change in kinematic direction, from NW-SE to E-W, started at C18 and was found in all basins of the Northeast Atlantic Ocean. This evolution was well marked by the change in direction of the transform faults that limited the seafloor basins of the Northeast Atlantic Ocean, such as the Jan Mayen faults (Figures 3.9 and 3.16). It preceded the end of oceanic accretion in the Labrador–Baffin axis at C13 (Oligocene, 33.9 My).

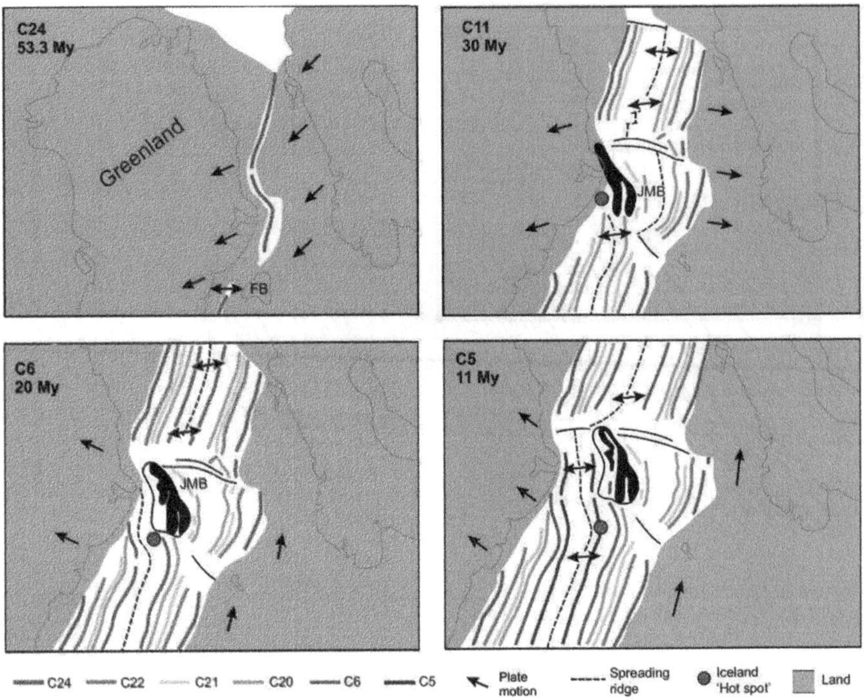

Figure 3.16. *Kinematic reconstructions of the Northeast Atlantic opening and individualization of the Jan Mayen Block or Microcontinent (JMB) (from Ellis and Stocker 2014). FB: Faroe Bank*

From C6 (Burdigalian, 20 My) (Ogg *et al.* 2006) to the present, the pattern of magnetic anomalies changed radically at the Reykjanes Ridge (Figure 3.15). They were no longer segmented or orthogonal to the kinematic direction and became again parallel to anomalies 24–18, although the oceanic opening direction remained oblique.

They are expressed over a major bathymetric anomaly, defining a morphological V in the southwest of Iceland (Figure 3.17). We will return to the interpretations of this V-shape of the ridge in section 3.2.5.3. The process of oblique oceanic expansion observed at Reykjanes Ridge is close to that described much further north at Mohns Ridge (Dauteuil and Brun 1993).

While the magnetic anomalies were well defined in the south and in the center of the Reykjanes Basin as early as C24, this was not the case north of the basin for the oldest anomalies, notably during C24 and C23 (Figures 3.9, 3.13 and 3.14). This could suggest a propagation of the continental breakup, and thus of oceanization, from south to north. This northward propagation of the oceanic rift was synchronous with the southward propagation of the Ægir rift (section 3.2.4.5). This dynamic ended with the separation of a continental block from Greenland, Jan Mayen's future microcontinent (section 3.2.4.6; Figure 3.16).

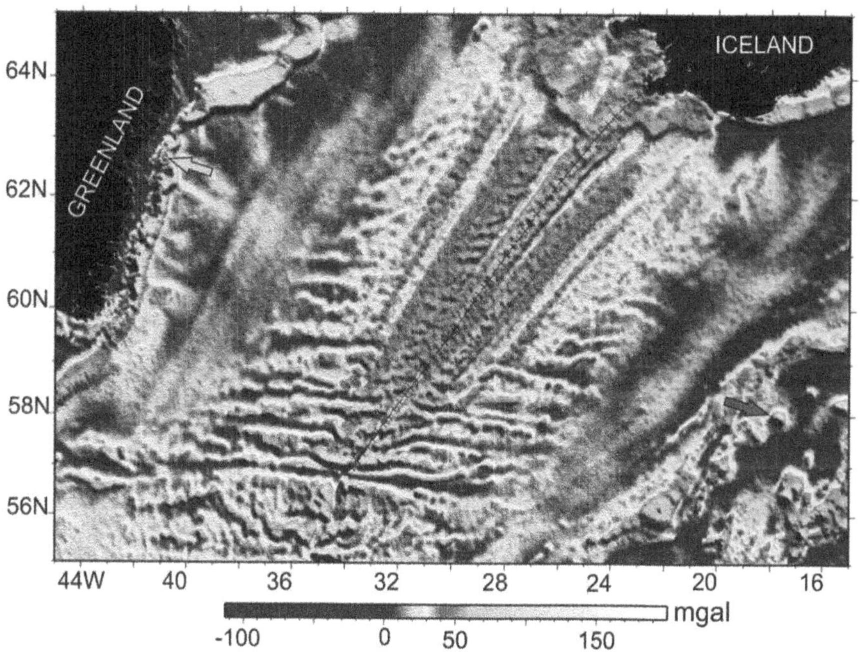

Figure 3.17. *Map of open-air gravimetric anomalies of the Reykjanes Basin (from Martinez* et al. *2019)*

COMMENT ON FIGURE 3.17.– *Note the E-W direction of the transform faults in the basin (from C18 onwards) and their obliquity relative to the ridge. Note also the V-shape of the Reykjanes Ridge. The red arrow (right) corresponds to one of the Paleocene-Eocene magmatic centers in the realm covered by the traps and the yellow arrow to one of the Eocene magmatic centers of the volcanic passive margin east of*

Greenland (section 3.4.5.3). These centers, which can also be studied on land (such as those of Skye or Mull in the British Isles), form seamounts at sea with very dense and deep crustal roots, both in the intraplate domain and at the level of the passive volcanic margins (Bott and Tuson 1973; Callot and Geoffroy 2004).

3.2.4.4. *The GIFR ridge*

A striking fact is the apparent disappearance of magnetic anomalies between Greenland and the Faroe Islands at the latitude of Iceland (Figure 3.14). This zone is abnormally shallow (about 600 m maximum) and is associated with a very thick crust (Figure 3.20). It constitutes the Greenland–Faroes–Iceland aseismic Ridge (GIFR) (Figure 1.1) (Hjartarson *et al.* 2017). Magnetic anomalies with diffuse contours exist through the GIFR (Figure 3.18), but cannot be easily correlated with anomalies in the ocean basins located south and north of the ridge. Some authors propose that the GIFR is bounded northwest of the Faroe Islands by a transform fault (Figure 3.9). However, this fault is not clearly visible either in bathymetry (or open-air gravimetric anomalies) or in magnetic maps (Figure 3.18). We will come back to the definition of the Greenland–Faroes–Iceland Ridge in section 3.3.3.

3.2.4.5. *The extinct Ægir Ridge*

North of the GIFR is a complex segment of the Northeast Atlantic characterized by the successive opening of two distinct ocean basins (Figures 3.7, 3.9, 3.13 and 3.14).

From C24 to C13, the Northeast Atlantic did not open in the continuity of the Reykjanes Ridge but along an axis located in the east in the Norwegian Sea (Ægir axis). Magnetic anomalies in this basin suggest that the continental breakup and oceanic expansion have propagated from north to south, i.e. toward the GIFR (Gernigon *et al.* 2015, 2019). The Ægir Ridge, now inactive (and therefore subsiding) and whose paleo-rift is associated with a strong negative open-air gravimetric anomaly (Figure 3.7), is a former expansion axis with an apparent curved shape whose connection to the Reykjanes Ridge (Figure 3.12) is unclear (fault or lack of connection?). The nature of the passive margins of the Ægir basin in the Norwegian Sea is volcanic (Figures 3.9 and 3.13). The Ægir basin is bounded to the north by the eastern Jan Mayen transform fault, which has a curved shape (NNW-SSE direction in the east and NW-SE direction in the west) related to the kinematic change in the opening of the Northeast Atlantic at C18 (section 3.2.4.3).

Oceanic accretion in the Ægir basin ended at the Oligocene (around C13), as shown by the last magnetic anomaly expressed in this basin (Gernigon *et al.* 2015). This gradual extinction was preceded by a continental extension into the continental lithosphere east of Greenland (Figure 3.16). This extension led to a continental breakup east of Greenland at the end of the Oligocene (C7-C6; Figure 3.16) and the establishment of a new axis of oceanic accretion, the Kolbeinsey Ridge, connected to the Reykjanes Ridge west of proto-Iceland (Figures 3.13 and 3.16).

3.2.4.6. *The Jan Mayen microcontinent and the Kolbeinsey Basin*

This transfer of divergence to Greenland resulted in the isolation of a continental area that is now submerged, the Jan Mayen microcontinent (Figures 3.14 and 3.16).

The Jan Mayen microcontinent is thus bordered by two passive margins (Figure 3.9): to the east, the passive volcanic margin conjugated with that of Møre in Norway, and to the west, the passive margin of the Kolbeinsey basin, conjugated with the central-eastern Greenland margin. This microcontinent corresponds to a continental lithosphere that is particularly deformed and complex in structure because it was initially affected by Mesozoic rifting episodes. Then, it was reactivated, to the east and west, by the diachronous extension associated, respectively, with the margins of the Ægir and Kolbeinsey basins.

It is interesting to note again the interdependence between the accretionary system of the Labrador–Baffin axis and that of the northeast Atlantic. Indeed, the progressive westward jump in the axis of divergence between the Ægir and Kolbeinsey basins from C13 is synchronous with the cessation of the divergence between Canada and Greenland (section 3.2.2).

The Kolbeinsey Basin is bordered to the north by the west Jan Mayen Fault, oriented WNW-ESE, an opening direction from C18 (Figure 3.9). The active volcanic island of Jan Mayen is located close to this fault.

It is interesting to note that the Mohns Ridge, located north of the Jan Mayen transform faults, is a strongly oblique ridge relative to the kinematic direction, like the Reykjanes Ridge since C18 (Dauteuil and Brun 1993).

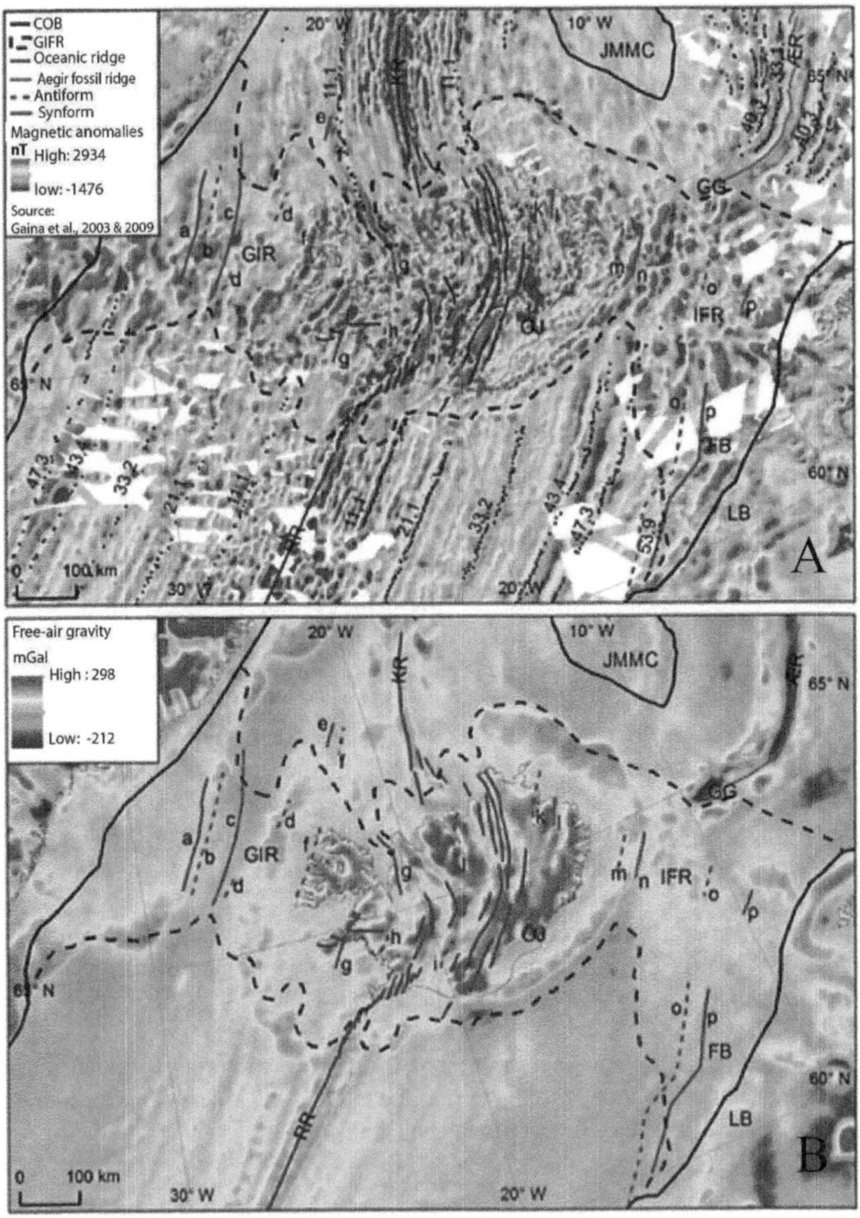

Figure 3.18. *Anomalies around Iceland (A) magnetic anomalies and (B) free-air gravity, along the Greenland–Iceland–Faroe aseismic ridge (based on Hjartason* et al. *2017)*

COMMENT ON FIGURE 3.18.– *JMMC: Jan Mayen Microcontinent, GIR: Greenland–Iceland Ridge, IFR: Iceland–Faroe Ridge, ÆR: Ægir Extinct Ridge, GG: Gridar Gorge, FB: Faroe Basin, LB: Lousy Bank, OJ: Öræfajökull. Note the disappearance of linear magnetic anomalies along the IFR and GIR aseismic ridges. In the north of Iceland, the Kolbeinsey Ridge (KR) is connected to the north of the active Icelandic rift by the Tjörnes transform zone. The Icelandic Rift is connected to the Reykjanes Ridge (RR) by the South Iceland Seismic Zone and the Reykjanes Peninsula.*

3.2.5. *The Northeast Atlantic region: mantle plume or not?*

Thulean magmatism (Figures 3.11(a) and 3.13) is classically attributed to the effect of a deep mantle plume (Richards *et al.* 1989; White and McKenzie 1989; Kerr 1995). The head of this mantle plume from the D'' layer is thought to have spread below the base of the northeast Atlantic lithosphere in the Paleocene. Its residual effect, the Icelandic hot spot, is thought to have continued until the present (Figures 3.5 and 1.6). The supporters of this hypothesis base themselves, among other arguments, on the regional evidences previously presented in this chapter. Notably, the Late Cretaceous to Danian regional uplift, preceding the Thulean basalt emission, would correspond to the impact of the head of a plume beneath the very heterogeneous lithosphere, in terms of thermal thickness, of the Northeast Atlantic domain. The north Atlantic geoid high (Figure 3.1(b)) is frequently related to a very broad deep-sourced mantle upwelling beneath Iceland.

The hypothesis of the existence of a "Thulean mantle plume" is the subject of scientific controversy, which we will present in this chapter.

3.2.5.1. *The Icelandic hot spot*

Seismic tomography shows unambiguously the existence in the upper mantle of a slow, and therefore probably hot, zone beneath Iceland (Wolfe *et al.* 1997) (Figures 1.6 and 3.19). Iceland is today unanimously considered by geophysicists as a hot spot. Since the 1990s, the accuracy of tomographic images has been improved with the deployment of broadband seismometer networks (Bjarnason *et al.* 1996; Wolfe *et al.* 1997; Foulger *et al.* 2000, 2001; Montagner and Ritsema 2001; Allen *et al.* 2002). They have made possible the illustration of this anomaly in 3D. The predominant structure, proposed by the different authors, is that of a sub-cylindrical seismic velocity anomaly under Iceland to a depth of 400 km (Figure 1.6). The thermal maximum would be under the Hofsjökull and Vatnajökull glaciers (Wolfe *et al.* 1997). Bijwaard and Spakman (1999), Monticelli *et al.* (2006) and He *et al.* (2015) modeled a thermal anomaly in the form of a mantle plume with a head about 200 km in diameter, an axis ascending under Iceland and a broad root, at the level of the interface between the core and the mantle. Nevertheless, the extension of this thermal anomaly to the lower mantle remains highly controversial, as seismic tomography interpretations are particularly equivocal beyond 400–600 km in the Northeast Atlantic (Ritsema *et al.* 1999; Foulger *et al.* 2000, 2001; Hosseini *et al.* 2018) (Figure 3.19).

3.2.5.2. *The mantle plume hypothesis*

While the existence of a current narrow hot spot under Iceland, leading to abnormal mantle melting in a diverging geodynamic system, is not contestable, linking this hot spot to the Paleogene magmatism of the Thulean Province (section 3.2.3) remains controversial, as other processes may explain the geological observables.

The Thulean plume hypothesis explains, for its supporters, the following observations, already presented in this chapter:

– The existence of a Late Cretaceous to Paleocene regional uplift (section 3.2.3.1) with detrital influxes into the peripheral sedimentary basins (White and Lovell 1997); this uplift could be associated with an early regional mantle melting with magmatic rocks "underplating" beneath the crust, resulting in isostatic uplift (White and Lovell 1997) and/or the dynamic effect of mantle plume head establishment (Figure 3.5(a)).

– The suddenness and apparent simultaneity of magma spreading in the Lower Paleocene over an extremely large area from West Greenland to the British Isles may suggest the melting of a mantle plume head of considerable diameter, close to 2,000 km (section 3.2.3.1, Figures 3.5(a) and 3.11(a)). Magma production rates of basalts in the Paleocene and Eocene, before and during the Paleocene-Eocene continental breakup (passive volcanic margins), were significant (>2 km^3/year on a magmatic province scale; Coffin and Eldholm 1994) and the helium isotopic geochemistry of basalts suggests a deep mantle origin (Marty *et al.* 1998; Kent *et al.* 2004).

– Another argument of the proponents for an ascending mantle flow is the existence of the significant bathymetric anomaly along the Reykjanes Ridge, south of Iceland, between C6 and the present, forming a "V" structure pointing southward (section 3.2.4.3, Figure 3.17). This axial structure in bathymetric charts was for White and Lovell (1997) the illustration of a channeling of the mantle flow, considered as pulsatile, of the Icelandic hot spot under the Reykjanes Ridge.

– The oceanic crust is unusually thick in the ocean basins of the Northeast Atlantic (often more than 10 km) and especially along the GIFR and in Iceland

(Figure 3.20). Anomalies in oceanic crustal thickness are interpreted as adiabatic decompression of a mantle warmer than the one at the source of MORB (White and McKenzie 1989). The spectacular thickness of the GIFR crust is thus classically interpreted as the trace of the opening of the Atlantic above the residual tail of the Thulean plume, i.e. above the Icelandic hot spot (Figure 3.5(b)).

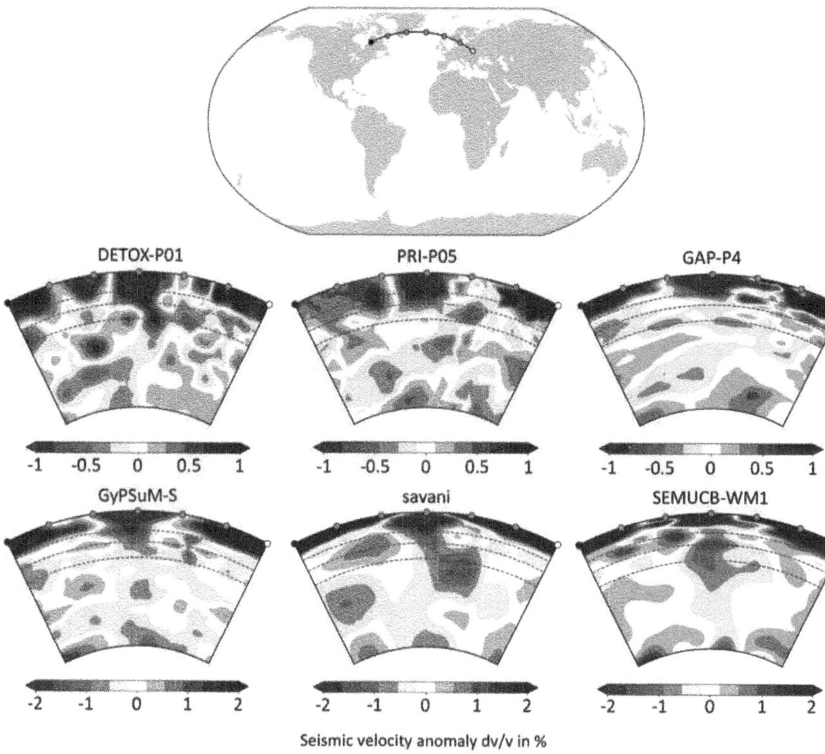

Figure 3.19. *Tomographic models following the same section of the Northeast Atlantic at the level of Iceland*

COMMENT ON FIGURE 3.19.– *The three levels of depths with dashed lines correspond to 410, 660 and 1,000 km. It is noted that while a seismic velocity anomaly exists in all models in the upper mantle under Iceland down to 400 km, the image in the deeper mantle differs from one model to another and is therefore not conclusive regarding the extension of the anomaly to greater depths (Hosseini* et al. *2018).*

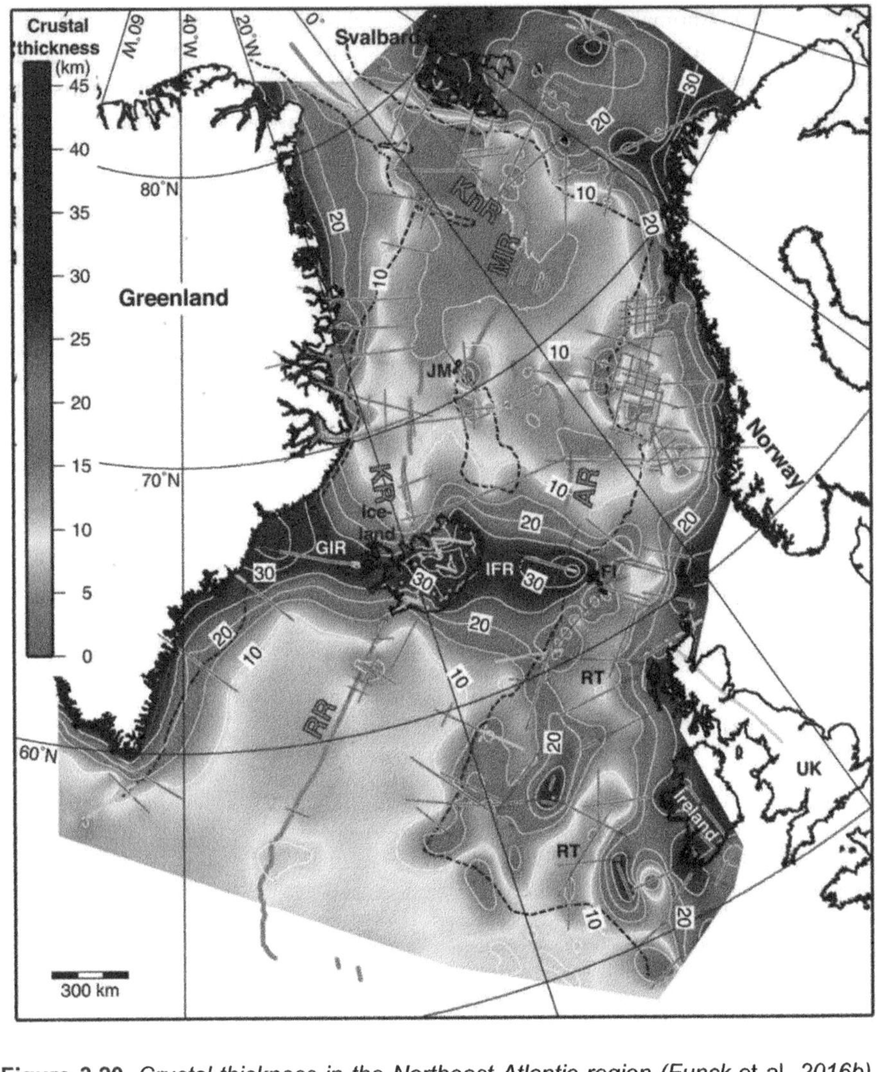

Figure 3.20. *Crustal thickness in the Northeast Atlantic region (Funck* et al. *2016b). RR: Reykjanes Ridge; KR: Kolbeinsey Ridge; KnR: Knipovich Ridge; AR: Ægir Ridge; MR: Mohns Ridge; RT: Rockall Trench; F: Faroe Islands; JM: Jan Mayen; GIR: Greenland–Iceland Ridge; IFR: Iceland–Faroe Ridge*

– The very magnesian volcanic formations (picrites) of Paleocene age that outcrop on either side of Greenland have been interpreted as being associated with a mantle that is much warmer than the upper mantle under the oceanic ridges. However, recent studies show that the potential temperature of the mantle during the emplacement of these magmas did not exceed 1,500°C, which corresponds to a thermal anomaly of less than 150°C under the lithosphere, before the latter was thinned by the formation of passive volcanic margins (Agranier *et al.* 2019; Hole and Natland 2019).

– From a general point of view, the geochemistry of trace elements and the isotopic geochemistry of the magmas of the Thulean province show a mantle signature enriched in incompatible elements, consistent with their derivation, from the beginning of magmatism in the Paleocene to the present (Iceland), from a common enriched chemical reservoir.

3.2.5.3. *Arguments against the mantle plume model*

The mantle plume concept, as defined in section 3.2.1, must fit the geological observables in the Northeast Atlantic region, and not the other way around. Theoretically, this model implies a succession of phenomena such as (i) a deep plume of large diameter origin, (ii) incubation, thermal erosion of the base of the lithosphere associated with regional mantle melting, (iii) continental extension and then continental breakup, (iv) continental and then oceanic lithosphere drift above the plume or residual hot spot, and finally (v) some dynamic effect of the current hot spot on the lithospheric bulge in the North Atlantic.

A number of observations do not fit the mantle plume hypothesis as follows:

– The Northeast Atlantic domain was globally in divergence and extension well before the establishment of Paleocene traps (section 3.2.2). It is as if the Cenozoic mantle melting anomaly was a terminal consequence of the different stages of Laurussia dislocation since the Devonian and not the cause of the divergence and oceanization. In other words, on a large scale in terms of time and space, the melting seems to be primarily the indirect or direct consequence of a regional continental extension. This is closer to a process of "passive" rather than "active" divergence, in the sense defined by Sengör and Burke (1978): the plume is not the consequence of lithospheric divergence, and it is this plate divergence that causes the regional melting attributed to the plume. Alternatively, the diverging zones would behave as attractors of mantle plumes.

– The dating of basalts at the scale of the magmatic province is not consistent with a simple model of progressive spreading of a plume head under the lithosphere (Figure 3.5(a)). During the Paleocene, the most recent kinematic models place the rising axis of the plume (the Icelandic hot spot) at the level of eastern Greenland (Figures 3.5(a) and 3.21). However, mantle melting did not begin in this zone (which includes the Faroe Islands) until C26, i.e. much later than at the extremities of the assumed plume (British Isles, West Greenland), where it began at C28N and C27R (Figure 3.11(b)).

– The distribution of Paleocene magmatism in the Northeast Atlantic region was far from being homogeneous in time and space (Figures 3.11(a) and 3.13). It was strongly dependent on the thermal (lithospheric thickness) and structural (faults, shear zones) inheritance of the continental lithosphere before the Cenozoic. At first order, volcanic formations were essentially located along Mesozoic rifts (Thompson and Gibson 1991) (Figures 3.9, 3.10 and 3.11(a)), unlike, for example, the magmatic provinces of the Deccan in India or the Parana-Etendeka. This overall relationship between magmatism, continental extension and fracture zones, whether ancient (inheritance) or syn-magmatic (Agranier *et al.* 2019), is evident. Mapping and statistical study of the vectors of magmatic flow in the dyke swarms that fed the basalts of the magmatic province show that these flows are lateral and that the magma comes from major magmatic centers that punctuate the Thulean Province (Callot 2004; Geoffroy *et al.* 2007) (Figures 3.11(a) and 3.17). The number of magmatic centers responsible for feeding basalts in the Thulean Province is of the order of 50, with a very heterogeneous distribution. Their spacing seems to depend primarily on the thickness of the thermal lithosphere (Figure 3.11(a)) (Callot 2004; Geoffroy *et al.* 2007), which is itself dependent on its age and tectonic history. In the British Isles, the distribution of paleogene magmatic centers (Skye, Mull, Arran, etc.) was rather directly controlled by faults, Late Caledonian (e.g. Great Glen Fault and Mull magmatic center) and Mesozoic in age (e.g. Camasunary Fault and the magmatic centers of Skye and Rùm). The relationship between the spacing of the magmatic centers and the thermal thickness of the lithosphere suggests that these magmatic centers, at the origin of Thulean volcanism, are located over small convection cells, themselves occurring within the transition zone between the conductive lithosphere and the convective upper mantle (Davaille and Jaupart 1994; Morency *et al.* 2002; Callot 2004; Geoffroy *et al.* 2007). The magmatic province of the Northeast Atlantic should therefore be viewed as a set of melting points related to small-scale sub-lithospheric convection with horizontal distribution of magma in the crust (Callot 2004; Geoffroy 2005; Geoffroy *et al.* 2007), and not as a homogeneously melting plume head. This does not exclude the plume hypothesis, as small-scale convection can occur in the plume head (Fleitout *et al.* 1986). However, other processes related to plate tectonics, some of them poorly explored, could lead to a generalized convective destabilization of the base of the lithosphere and a magmatic province. In particular, many studies show that small-scale convection induced by continental extension can explain the abundant magmatism of the passive volcanic margins without thermal anomalies in the asthenosphere (Keen and Boutilier 1995). The fact that the maximum magma in the Thulean province is on either side of the Greenland craton may also be explained by local convection induced by delamination effects at the edge of the Greenland craton (King and

Anderson 1998). Recent numerical models (Petersen *et al.* 2018) show that such top-down delamination (Anderson 2001; Morancy and Doin 2004) could have effects down to the deep mantle and be at the origin of a hot spot like Iceland.

– The bathymetric V south of Iceland (Figure 3.17) is not associated with a particularly thick oceanic crust, hence a higher rate of mantle melting, as would be expected if warm mantle fluxes were periodically injected under the axis of the ridge from the Icelandic hot spot. Other explanations have been proposed to account for this observation, such as that of a simple oceanic propagator (Hey *et al.* 2010; Benediktsdóttir *et al.* 2012) or that of a propagation along the ridge of small-scale convective cells (Martinez *et al.* 2019).

– The idea that the GIFR aseismic ridge, which has considerable crustal thickness (Figure 3.20), may correspond to the expression of an opening of the Atlantic above the Icelandic hot spot (Figure 3.5(b)) is highly debatable. This important point, which is directly related to the origin of Iceland, will be discussed in section 3.3.3.

From a more general point of view, there are multiple models that can explain a regional or local temperature anomaly in the upper mantle, without evoking a mantle plume of deep origin spreading under a lithosphere while thermally eroding it (Richards *et al.* 1989).

Supercontinents, surrounded by subducting zones such as Laurussia was (Figure 3.2), are excellent candidates to insulate an increasingly hot upper mantle, susceptible to gradually thermally erode the base of the lithosphere and generate intraplate magmatism without lithospheric extension. This process is well known and modeled (O'Neill *et al.* 2009). For example, the different "plumes" at the origin of the Phanerozoic magmatic provinces of Gondwana could have corresponded to a common upper mantle, isolated by subducting edges around the supercontinent, and gradually cooled and depleted by the extraction of magma from each magmatic province (Ganne *et al.* 2016).

As highlighted above, it has long been demonstrated (Keen and Boutilier 1995; Simon *et al.* 2009) that the existence of small-scale mantle convection during lithospheric breakup processes is sufficient to explain the rates of significant magma production observed at the passive volcanic margins. It is not necessary to invoke an active mantle that is significantly warmer than normal as in the mantle plume models (White and McKenzie 1989).

The high helium isotope ratios of basalts in the Northeast Atlantic region, both at the margins and in Iceland, are often used as an indisputable evidence of mantle melting of deep origin, with gravitational instability originating at the mantle/core boundary (Marty *et al.* 1998; Class and Goldstein 2005). The mantle involved in the melting is enriched in $^3He/^4He$ relative to the atmosphere up to a factor of 20. The usual interpretation of most geochemists is that of a partial melting of the Earth's deep "primitive" deep mantle (Moreira 2013; Graham *et al.* 2016). However, this common interpretation has been debated by several authors (Anderson 2001; Parman *et al.* 2005).

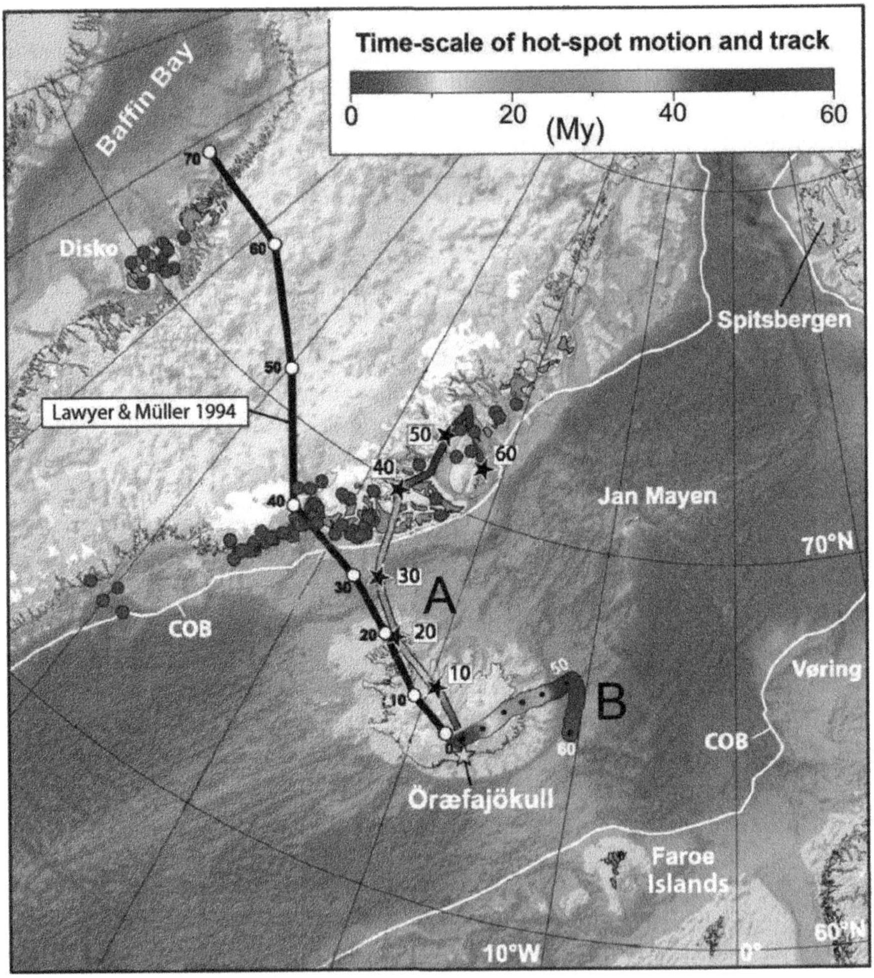

Figure 3.21. *Successive positions of the Icelandic hot spot during the drift of the Greenland plate. Simplified from Torsvik* et al. *(2015). Color plot: (A) position relative to the hot spot; (B) drift from the hot spot; numbers along the plots give the age in My. In black: model by Lawver and Müller (1994). COB: Continent-Ocean Boundary*

To conclude this discussion, it is therefore not excluded that the abundant Paleogene magmatism of the Northeast Atlantic region is above all linked to the adiabatic melting of a passive and probably slightly warmer mantle. This is in direct relation to the extension/dismantling of a large continent, Laurussia (Figure 3.2), made up of lithospheric elements of very varied ages and therefore thermal thicknesses. The notions of lithosphere thermal and structural inheritance are certainly crucial in understanding the processes that led to the opening of the Northeast Atlantic and the formation of Iceland (Schieffer *et al.* 2019).

3.3. The origin of Iceland

3.3.1. *The anomalous crust of the GIFR ridge and the deep structure of Iceland*

In the Northeast Atlantic Ocean, the most significant crustal thickness anomaly corresponds to the GIFR aseismic ridge that connects the east coast of central-eastern Greenland to the Faroe Islands archipelago including Iceland (Figures 1.1, 3.20 and 3.22). The crust has an average thickness of about 30 km, i.e. four to five times that of a standard oceanic crust (about 6–7 km).

A remarkable fact is the presence of SDRs along this ridge, in the continuity of the volcanic margins of central-eastern Greenland and the Faroe Islands (Figures 3.13 and 3.12(f)). They are locally discernible despite the poor quality of the seismic reflection data along this ridge. Along the Greenland–Iceland Ridge, the SDRs dip toward the east (toward Iceland) (Larsen and Jakobsdóttir, 1988), and along the Faroe-Iceland Ridge, they dominantly dip, when observed, toward the west (toward Iceland) (Hjartason *et al.* 2017) as those from the conjugated passive volcanic margins of Greenland and the Faroe Islands, located at the outer tips of the ridge (Figure 3.13). Where the SDRs are not discernible, a crust with a roof characterized by horizontal lava flows, and which is analogical in thickness, structure and seismic velocity to the FLF-type crust at the end of the volcanic passive margins (section 3.2.1), is found.

Figure 3.23(a) shows the depth of the base of the Icelandic crust, and Figure 3.23(b) shows the simplified geological map of the island with the age and dip direction of the basalts. In addition to the considerable thickness of the crust (up to 40 km), strong variations can be seen in crustal thickness without being able to be clearly correlated with the age of the crust. Thickness *minima* are located in the north and the south of the island. The age of the crust, or more rigorously the age of lavas, increases globally on either side of the active accretionary axis toward the eastern and western parts of Iceland.

The Icelandic crust is composed of an upper part dominated by lava and intrusions, and a lower part of a more enigmatic nature (Figure 3.24). The upper crust is of relatively constant thickness (about 7 km) (Foulger *et al.* 2003). It is seismogenic, and therefore with fragile behavior, and associated with very strong seismic velocity gradients. The seismic waves propagating in this crust show very little attenuation of a non-elastic nature. The lower crust is thick but very variable (15–30 km). It is characterized by high seismic wave velocities but varying very little with depth (of the order of 7–7.2 km/s) (Figure 3.24).

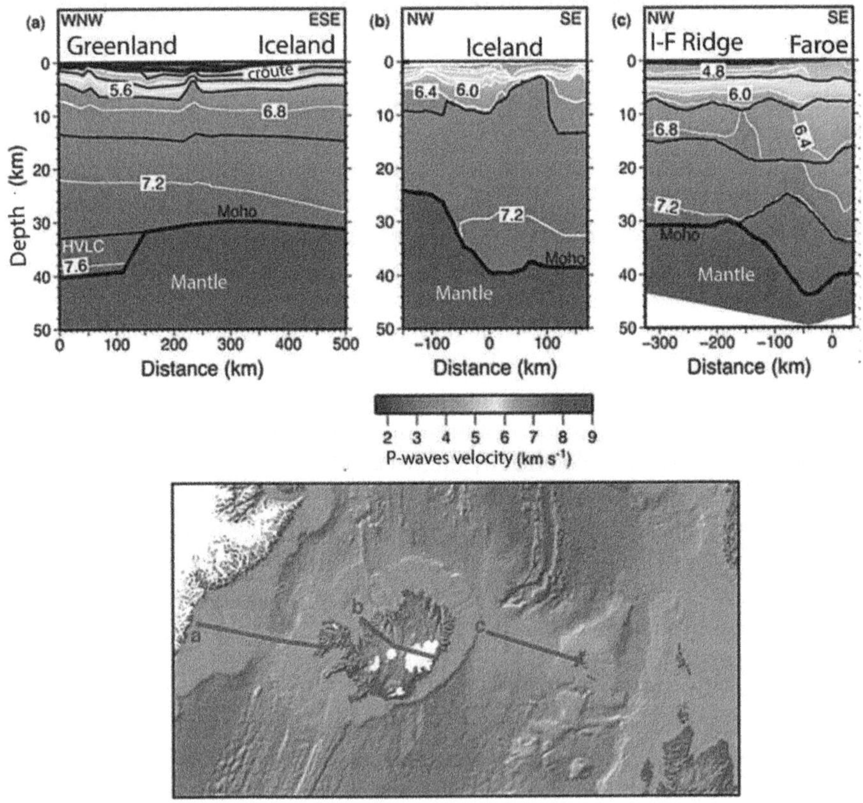

Figure 3.22. *Summary of the deep refraction seismic data along the GIFR ridge (from Funck* et al. *2016a). HVLC: high-velocity lower crust*

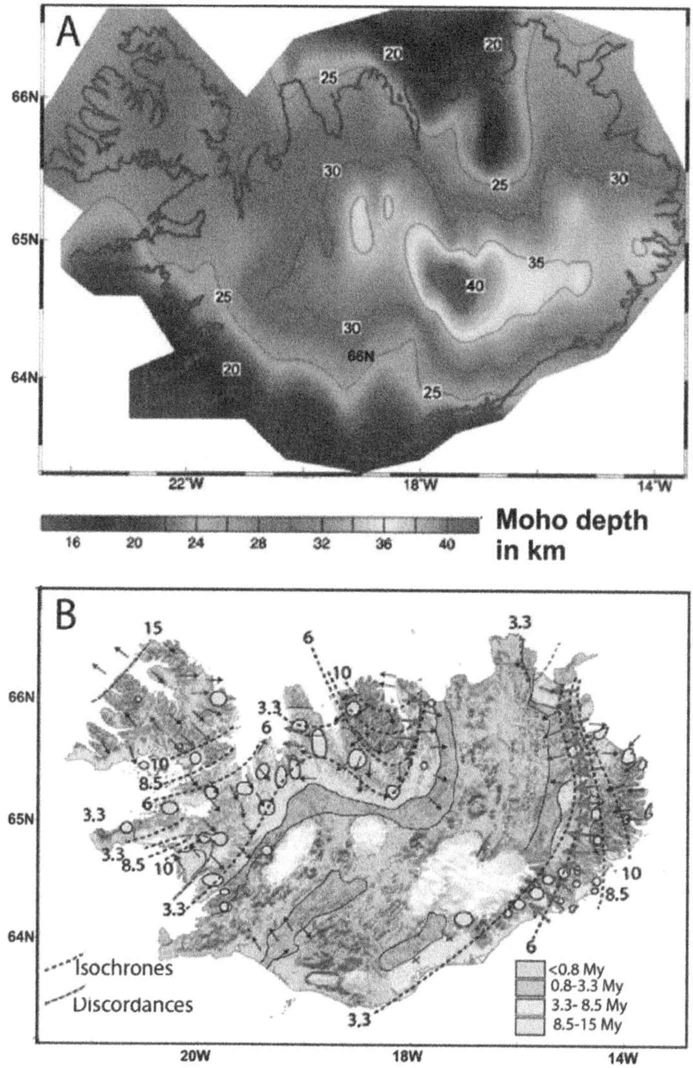

Figure 3.23. *Crustal structure of Iceland*

COMMENT ON FIGURE 3.23.– *(A) Moho isobaths in Iceland (from Darbyshire* et al. *2000). (B) Geological map of Iceland (after Bourgeois* et al. *1998; Johannesson and Sæmundsson 1998). The dip direction of the lavas, which form kilometric-thick prisms (SDRs) in Tertiary age formations, is indicated by the black (low dip angle) and red (steep dip) arrows.*

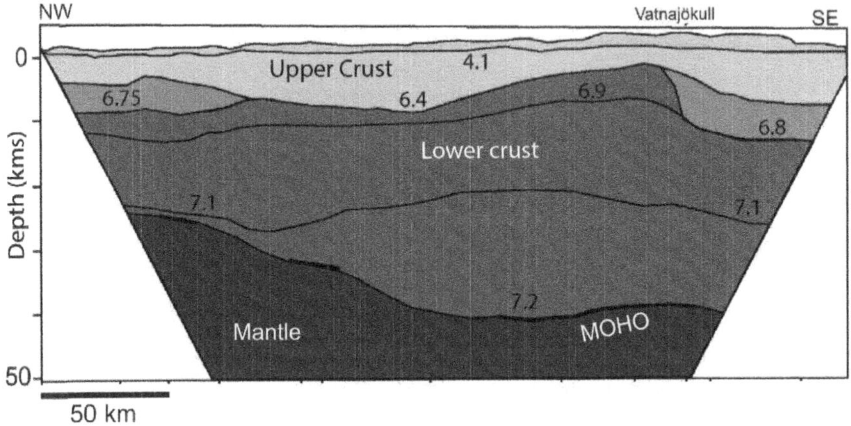

Figure 3.24. *Icelandic seismic crust. Seismic velocities are in km.s–1. Note the very low velocity gradients of the lower crust (variation of 0.8 km.s–1 over 30 km thick at the center of the island) (from Foulger* et al. *2003)*

COMMENT ON FIGURE 3.24.– *The seismic velocities along the GIFR are similar with those of volcanic passive margins (Figure 3.3(b)).*

This lower crust was initially interpreted as composed partly of magma, from magnetotelluric data (Beblo and Björnsson 1980) and by linear extrapolation of very strong thermal gradients observed in boreholes. However, these gradients do not correspond to the geothermal gradient itself, but are rather related to the circulation of high-temperature hydrothermal fluids (Agustsson and Flovenz 2005).

In contrast to the "magmatic" model, seismic velocities suggest that the lower crust of Iceland is mafic in composition, relatively cold and rigid (Menke and Sparks 1995). The seismic discontinuity of the Moho (Figure 3.24) is always difficult to illustrate in Iceland: it does not give clear reflections and results in only a very slight increase in P-wave velocity (Foulger *et al.* 2003).

3.3.2. *Icelandic SDRs*

The active tectonics of Iceland is described in Chapter 2. The exposed volcanic formations in Iceland are less than 15 My old (C5, Figure 3.23(b)). These formations form relatively symmetrical magnetic anomalies on either side of the active rift (northern end of the Reykjanes Ridge). The local anomalies in the symmetry of the isochrones show instability in the position of the accretion axis over time, which resulted in rift jumps.

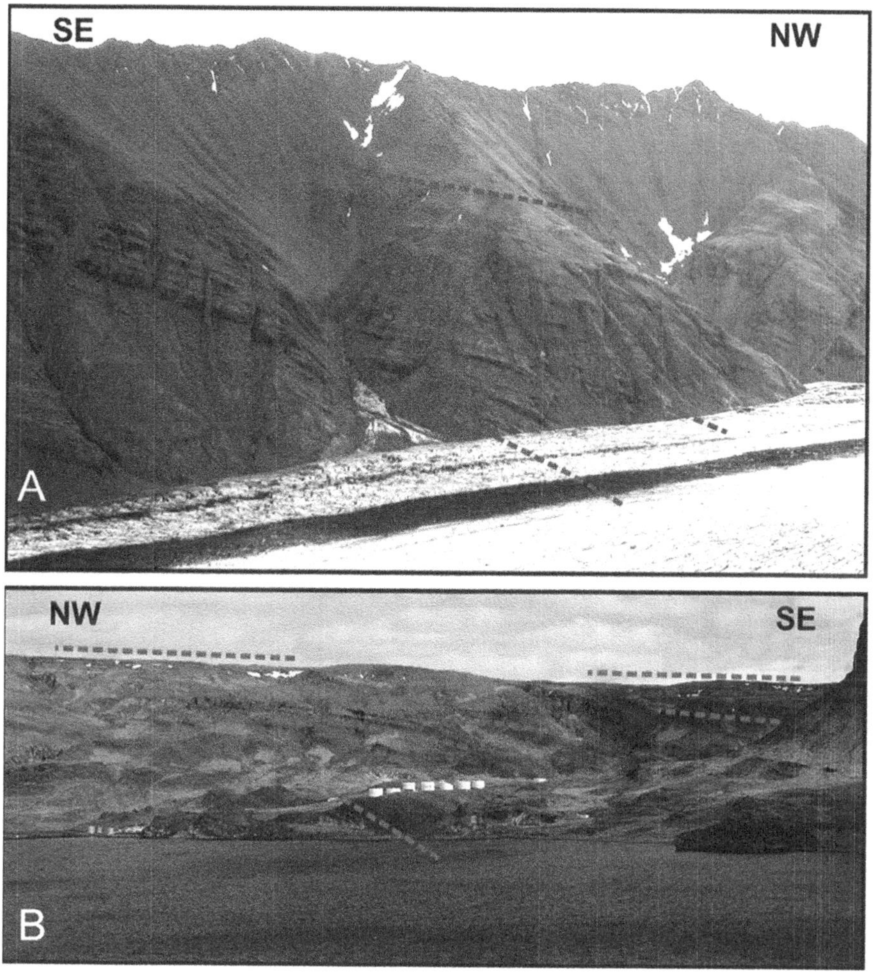

Figure 3.25. *Outcropping Icelandic SDRs. (A) SDRs of terminal Miocene age (about 6 My) plunging northwestwards in eastern Iceland (Höfn region). (B) SDRs of the same age, with opposite vergence, in western Iceland, Akranes region. Note the fan-shaped pattern of lavas with decreasing dip upward (red dashes) and the dip in both cases toward the active rift of Iceland (Laurent Geoffroy©)*

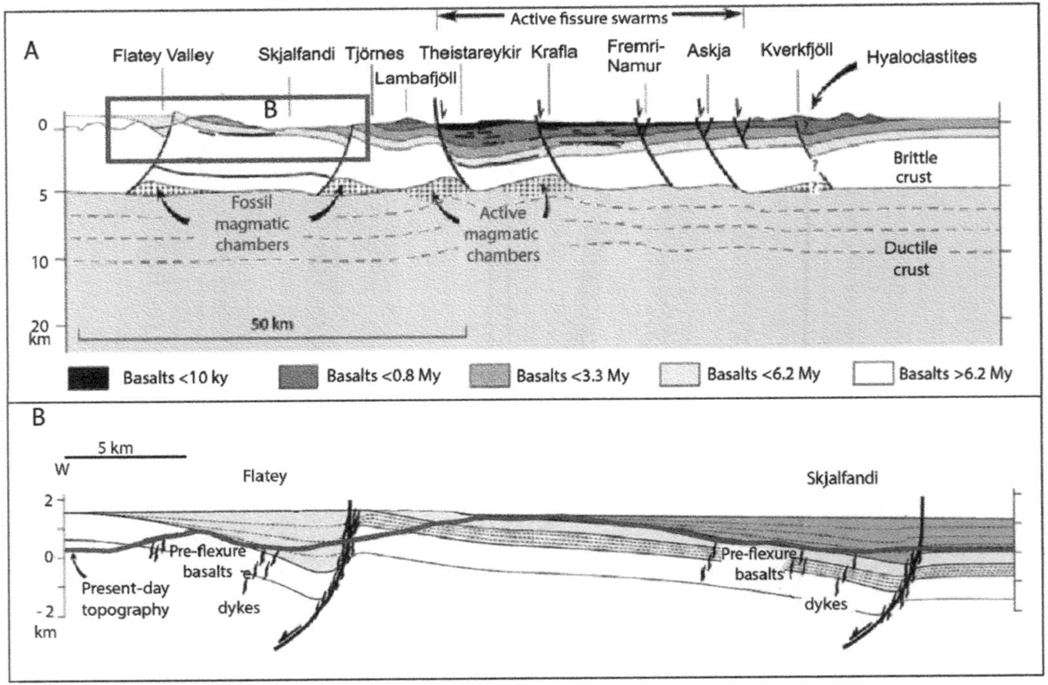

Figure 3.26. *Crustal structure in northern Iceland (modified from Bourgeois* et al. *2005).(A) Interpretive section through the Holocene fissure swarms of the North Volcanic Zone. (B) Interpretive section through the Flatey and Skjalfandi valleys (location in A)*

A crucial point in the geology of Iceland *stricto sensu* is the widespread existence of SDRs in the Miocene and Pliocene formations (Walker 1959) (Figures 3.23(b) and 3.25), as along the GIFR ridge in general and as at the adjacent volcanic continental passive margins (Figure 3.13).

Both east (Figure 3.25(a)) and west (Figure 3.25(b)) of Iceland, and independently of the structural complications associated with rift jumps (section 2.3.1.1), the Pliocene lavas are generally tilted toward the active rift (Figure 3.23(b)) by forming multi-kilometric prisms of unknown thickness. When observed in cross-section, it is shown that these lava piles were built in relation to faults with plurikilometric offsets (Bourgeois *et al.* 2005) dipping toward the continent (Greenland or Faroe), as at passive volcanic margins (Geoffroy 2005) (Figures 3.23(b) and 3.26). The fact that the Icelandic crust is composed of thick wedges of fan-shape lavas dipping toward the active rift (Figures 3.23 and 3.25) has long been recognized. Palmason (1980) developed a purely thermal and isostatic formation model, but it does not account for all observables such as prism edge faults and strong variations in lava dipping over short time periods (Figure 3.25), suggesting a tectonic control.

3.3.3. *Interpretations of GIFR and Iceland*

Two very different interpretations, discussed below, clash today regarding the origin of the GIFR and Iceland, especially to explain the structure and nature of their thick magmatic crust. One of these interpretations, the most classical and largely based on lava geochemistry, includes Iceland in the history of continental break-up and oceanic accretion above the Icelandic hot spot (Figures 3.5(b) and 3.21).

The other interpretation, which is still rudimentary, proposes that Iceland, like the GIFR ridge and part of the Northeast Atlantic oceanic domain, can be simply interpreted as the distal parts of passive volcanic margins developed above a very heterogeneous and chemically enriched mantle, and whose melting is directly related to the continental extension and kinematics of the plates.

This last interpretation is iconoclastic, because it not only does potentially reduce the area occupied by oceanic lithosphere in the Northeast Atlantic domain, but it also undermines the need for a regional thermal anomaly in the mantle to explain the geodynamics of this region. We will expose these two hypotheses from the facts that support them.

3.3.3.1. *The pattern of plate movement over a hot spot*

Aseismic bathymetric ridges, such as the Walvis-Rio Grande ridge in the South Atlantic (Graça *et al.* 2019) or the GIFR, characterized by a very thick crust, are considered generally as oceanic, and classically interpreted as related to an oceanic accretion above a hot spot. Most authors consider that hot spots are fixed, but that the oceanic accretionary plates and ridge may be mobile, in absolute motion, relative to the thermal anomaly location, and with possible relocations of the ridge above the hot spot (Figure 2.1) (Dyment *et al.* 2007 and references included). These relocations could explain the small-scale rift jumps in time and space in Iceland during the Neogene. This model assumes the persistence of a fixed upward mantle flow (hot spot) in the context of global mantle convection, whose geometry is however constantly changing over time in relation to plate tectonics (Coltice *et al.* 2017).

This model is difficult to apply to the entire GIFR, even when oceanic accretion (and hence relative plate motion) during global drift is taken into account (Figure 3.5). The symmetry of the magnetic anomalies observed at the Reykjanes Basin immediately south of the GIFR (Figures 3.9, 3.14 and 3.15) suggests that the GIFR (if interpreted as oceanic lithosphere) is associated with continuous oceanic accretion from C24 to the present day with no evidence of a shift of the ridge over a fixed hot spot that would permanently relocate it.

The apparent trajectory of the Icelandic hot spot under the North American plate considered fixed has been modeled using different methods, either by assuming that the Icelandic hot spot was fixed (Lawver and Muller 1994) or, more recently, by introducing an absolute shift of the hot spot during the Tertiary (Torsvik *et al.* 2015) (Figure 3.21). But the problem, despite this correction, remains unsolved. Those plate kinematics models suggest that during the Paleogene the hot spot was located beneath the Greenland basement. Only the Greenland–Iceland part of the GIFR could be explained, and this only between 35 My and the present for both models (Figures 3.5 and 3.21). The entire GIFR has no clear explanation in terms of plate and ridge drift above a hot spot.

3.3.3.2. *A continental axis stretched between Greenland and Europe?*

Three important points have been overlooked in the interpretation of GIFR and Iceland:

– The first point concerns the extraordinary coincidence between the GIFR and the West Caledonian front (Foulger *et al.* 2005) (Figure 3.9). This coincidence as well as new observations in the lithospheric mantle of northeast

Greenland led Schiffer *et al.* (2015) to suggest that part of an oceanic slab may have remained fossilized in the GIFR lithosphere during the divergence. According to Schiffer *et al.* (2015), the partial melting of this panel could explain not only the incompatible element-enriched chemistry of the lavas but also the abnormal rate of mantle melting along the GIFR (Figure 3.16).

– The second point is related to the fact that the crust composing the GIFR is structurally (SDRs and FLF-type crust; section 3.2.1) and physically (P-wave velocities in the upper and lower crust) very similar to that of a continental crust of volcanic passive margin (Foulger *et al.* 2019). The significant electrical conductivity of the lower crust of Iceland (Figure 3.24) is not related to the presence of magma (section 3.3.1). However, the lower continental crusts are characterized by high electrical conductivities, especially beyond 20 km depth, which are variously interpreted (Hyndman and Hyndman 1968; Hyndman *et al.* 1993; Yang 2011). It has recently been proposed, based on gravimetric inversions and geochemical data, that there is a continental crust under lava piles in Southeast Iceland (Torsvik *et al.* 2015). The existence of such non-oceanic material was already suggested by the incompatibility between the rate of oceanic accretion in Iceland and the width of the island, even taking into account rift-jumping (Foulger 2006).

– The third point is that no accretionary process at an active slow-spreading ridge, nor any ophiolite, demonstrates structural formation in the upper crust comparable to the SDRs observed in passive volcanic margins.

These different points can be combined speculatively to propose a different scenario for the GIFR formation.

In the construction processes of the passive volcanic margins, and due to the pure shear deformation of the lithosphere, a continental block, called block C, is isolated early and may be preserved during ongoing deformation and magmatic processes (Figure 3.27(a)). This block simply corresponds to the common footwall of the continent-dipping detached faults that border the inner SDR prisms of two conjugated passive margins. The genesis and evolution of this block C during the later stages of deformation and divergence were reproduced during a thermomechanical modeling of lithospheric deformation in relation to mantle convection (Geoffroy *et al.* 2015). These models show that the block tends to widen and thin with time due to extensional faulting (Figure 3.27(b)). Of course, since the initial stage of deformation, a large volume of magma is dyke-injected into the block C crust and extruded as SDR or horizontal lavas.

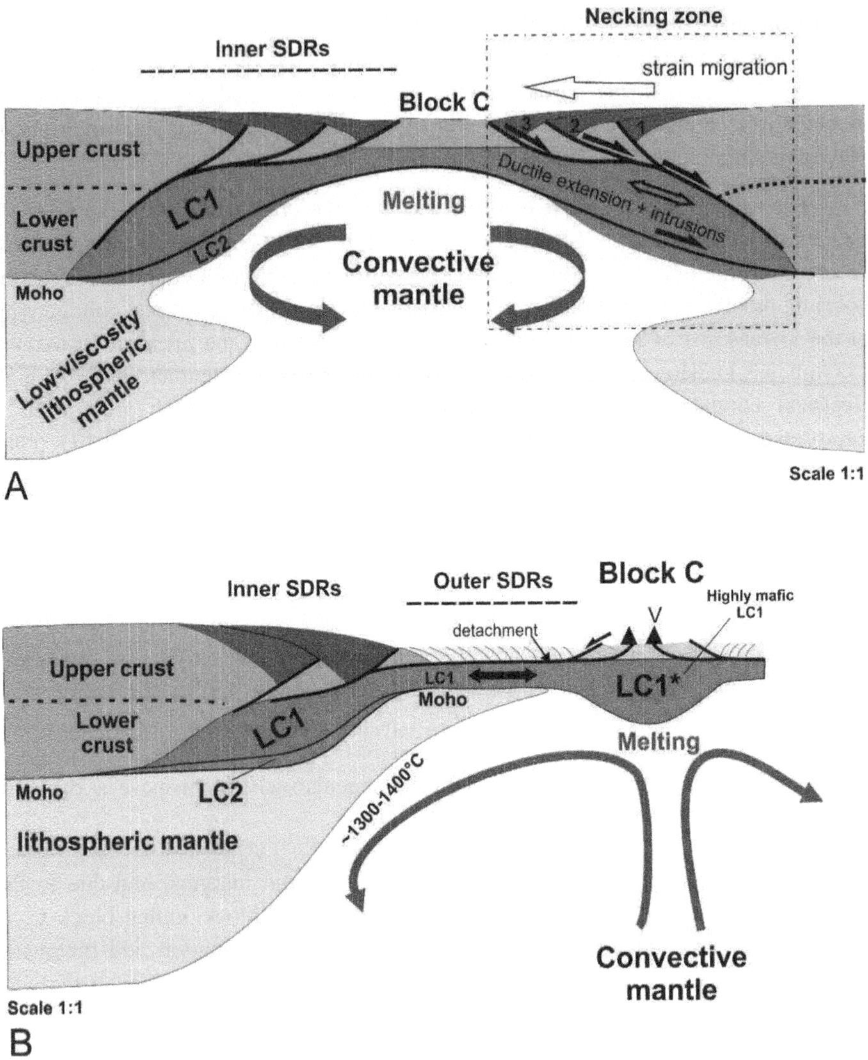

Figure 3.27. *(a and b) Schematic models of the formation of volcanic passive margins*

COMMENT ON FIGURE 3.27.– *(A) Block C at the initial stage of evolution of two conjugated passive volcanic margins during the formation of the inner SDRs. (B) Exhumation stage of the injected lower continental crust LC1 (section 3.2.1) below the external SDRs with distal isolation of block C, which is increasingly fragmented. V: polygenic volcanoes. This situation could correspond to the Greenland–Iceland*

Ridge (LC1 exhumation zone) and Iceland today (dilacerated block C with however thick lower continental crust) (according to Geoffroy et al. *2015, 2020).*

Recent observations on deep seismic reflection suggest that such a block C is present in the Laxmi volcanic rift off the coast of India, with similar geometry shown in Figure 3.27(a) for inner SDRs (Nemcok and Dunbar 2014; Geoffroy *et al.* 2020). These observations are important because this rift, although aborted, goes locally to the formation of the outer SDRs (Figures 3.3(b) and 3.27(b)) and to the breakup of block C, with the initiation of a new oceanic crust. These data also suggest that the outer SDRs are settling over a lower continental crust that is heavily injected with magma and exhumed.

This result, if confirmed, would be essential. It would demonstrate that the process of separation, in the case of magmatic breakups, is a process that preserves the ductile lower crust of the continents by detaching the upper crust (passive exhumation process).

These mechanisms are applicable to Iceland and the GIFR ridge (Figure 3.27(b)) (Foulger *et al.* 2019; Geoffroy *et al.* 2020).

In this perspective, and taking into account the Caledonian heritage, it is quite possible that the GIFR ridge consists of a ductile or semi-ductile lower continental crust extremely injected with magma (hence its thickness) and stretched, re-covered with syn-tectonic lavas, forming SDRs or sub-horizontal or slightly inclined lavas (FLF), located directly on the exhumed lower crust (Figure 3.27(b)). This crust of Caledonian origin would be relatively thick and ductile, since it is unaffected, because of its approximate E-W direction, by the Devonian orogenic collapse and the Mesozoic thinning stages (section 3.2.2.3). Iceland would then correspond to a residual block C (Figure 3.27(b)) with continental breakup and, possibly, recent oceanization, corresponding to the current active rift. A similar process would probably also apply to the Rio Grande aseismic ridge in the South Atlantic where evidence of open ocean continental crust has been discovered (Stica *et al.* 2013).

The consequences of such a general interpretation, associating the role of lithospheric inheritance to continental breakup processes over a warm mantle, go even further because the extent of SDRs in the Reykjanes Basin suggests that oceanic accretion in this basin may have started only from C22 and not from C24. Indeed, the seismic reflection profiles carried out during the SIGMA refraction seismic campaigns off the coast of Greenland have clearly demonstrated the existence of SDRs under magnetic anomalies 24 to 22, above the thinning zone of the continental crust, thus at the very level of the passive continental margin. This interpretation would significantly limit the seafloor domain in the Northeast Atlantic.

Of course, these new hypotheses must be confirmed. A recent study, consisting of the advanced interpretation of a SIGMA seismic refraction line across the Greenland–Iceland ridge, fully supports the occurrence of a continental crust beneath the basalts (Yuan *et al.* 2020).

In the coming years, those new ideas will need to be explored further in so far as they represent a significant advance in our understanding of oceanic realms.

3.4. References

Agranier, A., Maury, R.C., Geoffroy, L., Chauvet, F., Le Gall, B., Aviana, A. (2019). Volcanic record of continental thinning in Baffin Bay margins: Insights from Svartenhuk Halvø Peninsula basalts, West Greenland. *Lithos*, 334–335, 117–140.

Allen, R.M., Nolet, G., Morgan, W.J., Vogfjörð, K., .Bergsson, B.H., Erlendsson, P., Foulger, G.R., Jakobsdóttir, S., Julian, B.R., Pritchard, M., Ragnarsson, S., Stefánsson, R. (2002). Imaging the mantle beneath Iceland using integrated seismological techniques, *J. of Geophys. Res.*, 107(B12), 2325, doi: 10.1019/2001JB000595.

Anderson, D.L. (2001a). The helium paradoxes. *Proc. Nat. Acad. Sci.*, 95, 4822–4827.

Anderson, D.L. (2001b). Top-down tectonics? *Science*, 293, 2016–2018.

Annen, C. and Sparks, R.S.J. (2002). Effects of repetitive emplacement of basaltic intrusions on thermal evolution and melt generation in the crust. *Earth Planet. Sci. Lett.*, 203, 937–955.

Augland, L.E., Jones, M.T., Svensen, H.H., Planke, S. (2016). Improving the geochronology of the North Atlantic Igneous Province [Online]. Available at: www.largeigneousprovinces.org/16dec.

Beblo, M. and Björnsson, A. (1980). A model of electrical resistivity beneath NE-Iceland, correlation with temperature. *J. Geophys.*, 47, 184–190.

Benediktsdóttir, Á., Hey, R., Martinez, F., Höskuldsson, Á. (2012). Detailed tectonic evolution of the Reykjanes Ridge during the past 15 Ma. *Geochem. Geophys. Geosyst.*, 13 [Online]. Available at: http://doi.org/10.1029/2011GC003948.

Bijwaard, H. and Spakman, W. (1999). Tomographic evidence for a narrow whole mantle plume below Iceland. *Earth Planet. Sci. Lett.*, 166, 121–126.

Björnsson, A. (1985). Dynamics of crustal rifting in NE Iceland. *J. Geophys. Res.*, 90(B12), 10.151–10.162.

Bott, M.H.P. and Tuson, J. (1973). Deep structure beneath the Tertiary volcanic regions of Skye, Mull, and Ardnamurchan, North-west Scotland. *Nature*, 242, 114–116.

Bourgeois, O., Dauteuil, O., Van Vliet-Lanoë, B. (1998). Subglacial volcanism in Iceland: Tectonic implications. *Earth Planet. Sci. Lett.*, 164(1–2), 165–178.

Bourgeois, O., Dauteuil, O., Hallot, E. (2005). Rifting above a mantle plume: Structure and development of the Iceland Plateau. *Geodin. Acta*, 18(1), 1–22.

Callot, J.P. (2002). Origine, structure et développement des marges volcaniques : l'exemple du Groenland : interactions manteau-lithosphère en contexte de panache. PhD Thesis, Université Pierre et Marie Curie, Paris.

Callot, J.P. and Geoffroy, L. (2004). Magma flow in the East Greenland dyke swarm inferred from study of anisotropy of magnetic susceptibility: Magmatic growth of a volcanic margin. *Geophys. J. Int.*, 159(2), 816–830.

Campbell, I.H. (2005). Large igneous provinces and the mantle plume hypothesis. *Elements*, 1, 265–269.

Cazenave, A. and Feigl, K. (1994). *Formes et mouvements de la terre*. Belin, Paris.

Chalmers, J.A. and Laursen, K.H. (1995). Labrador Sea: The extent of continental and oceanic crust and the timing of the onset of seafloor spreading. *Mar. Petrol. Geol.*, 12, 205–217.

Chauvet, F., Geoffroy, L., Guillou, H., Maury, R.C., Le Gall, B., Agranier, A., Aviana, A. (2019). Eocene continental breakup in Baffin Bay. *Tectonophysics*, 757, 170–186.

Chian, D.P. and Louden, K.E. (1994). The continent-ocean crustal transition across the Southwest Greenland Margin. *J. Geophys. Res.*, 99, 9117–9135.

Chian, D., Louden, K.E., Reid, I. (1995). Crustal structure of the Labrador Sea conjugate margin and implications for the formation of nonvolcanic continental margins. *J. Geophys. Res.*, 100(B12), 24239–24253 [Online]. Available at: http://doi.org/10.1029/95JB02162.

Class, C. and Goldstein, S.L. (2005). Evolution of helium isotopes in the Earth's mantle. *Nature*, 436, 1107–1112.

Coffin, M.F. and Eldholm, O. (1994). Large igneous provinces: Crustal structure, dimensions, and external consequences. *Rev. Geophys.*, 32, 1–36.

Courtillot, V., Richards, M., Duncan, R. (1989). Flood basalts and hotspot tracks: Plume heads and tails. *Science*, 246(4926), 103–107.

Courtillot, V., Jaupart, C., Manighetti, I., Tapponnier, P., Besse, J. (1999). On causal links between flood basalts and continental breakup. *Earth Planet. Sci. Lett.*, 166, 177–195.

Courtillot, V., Davaille, A., Besse, J., Stock, J. (2003). Three distinct types of hotspots in the Earth's mantle. *Earth Planet Sci. Lett.*, 205, 295–308.

Dam, G., Larsen, M., Sørensen, J.C. (1998). Sedimentary response to mantle plumes: Implications from Paleocene onshore successions West and East Greenland. *Geology*, 26, 207–210.

Darbyshire, F.A., White, R.S., Priestley, K.F. (2000). Structure of the crust and uppermost mantle of Iceland from a combined seismic and gravity study. *Earth Planet. Sci. Lett.*, 181(3), 409–428.

Dauteuil, O. and Brun, J.P. (1993). Oblique rifting in a slow-spreading ridge. *Nature*, 361, 145–148.

Davaille, A. and Jaupart, C. (1994). Onset of thermal convection in fluids with temperature-dependent viscosity: Application to the oceanic mantle. *J. Geophys. Res.*, 99(B10), 19853–19866.

De Paor, D.G., Bradley, D.C., Eisenstadt, G., Phillips, S.M. (1989). The Arctic Eurekan orogen: A most unusual fold-and-thrust belt. *Geol. Soc. Am. Bull.*, 101(7), 952–967.

Dumoulin, C., Doin, M.P., Arcay, D., Fleitout, L. (2005). Onset of small-scale instabilities at the base of the lithosphere: Scaling laws and role of pre-existing lithospheric structures. *Geophys. J. Int.*, 160, 344–356 [Online]. Available at: http://doi.org/10.1111/j.1365-246X.2004.02475.xGJI.

Dyment, J., Lin, J., Baker, E. (2007). Ridge-hotspot interactions. What mid-ocean ridges tell us about deep earth processes. *Oceanography*, 20(1), 102–115.

Eldholm, O. and Grue, K. (1994). North Atlantic volcanic margins: Dimensions and production rates. *J. Geophys. Res.*, 99(B2), 2955–2968.

Ellis, D. and Stoker, M.S. (2014). The Faroe-Shetland basin: A regional perspective from the Paleocene to the present day and its relationship to the opening of the North Atlantic Ocean. In *Hydrocarbon Exploration to Exploitation West of Shetlands*, Cannon, S.J.C. and Ellis, D. (eds). Geological Society of London Special Publications, London [Online]. Available at: http://doi.org/10.1144/SP397.

Fleitout, L., Froidevaux, C., Yuen, D. (1986). Active lithospheric thinning. *Tectonophysics*, 132, 271–278.

Foulger, G.R. (2006). Older crust underlies Iceland. *Geophys. J. Int.*, 165, 672–676.

Foulger, G.R., Pritchard, M.J., Julian, B.R., Evans, J.R. (2000). The seismic anomaly beneath Iceland extends down to the mantle transition zone and no deeper. *Geophys. J. Inter.*, 142, F1–F5.

Foulger, G.R., Pritchard, M.J., Julian, B.R., Evans, J.R., Allen, R.M., Nolet, G., Morgan, W.J., Bergsson, B.H., Erlendsson, P., Jakobsdottir, S. *et al.* (2001). Seismic tomography shows that upwelling beneath Iceland is confined to the upper mantle. *Geophys. J. Inter.*, 146, 504–530.

Foulger, G.R., Du, Z., Julian, B.R. (2003). Icelandic-type crust. *Geophys. J.*, 155, 567–590.

Foulger, G.R., Natland, J.H., Anderson, D.L. (2005). A source for Icelandic magmas in remelted Iapetus crust. *J. Volcan. Geotherm. Res.*, 141, 23–44.

Foulger, G.R., Doré, T., Emeleus, C.H., Franke, D., Geoffroy, L., Gernigon, L., Hey, R., Holdsworth, R.E., Hole, M. *et al.* (2019). The Iceland microcontinent and a continental Greenland-Iceland-Faroe Ridge. *Earth Sci. Rev* [Online]. Available at: https://doi.org/10.1016/j.earscirev.2019.102926.

Funck, T., Erlendsson, Ö., Geissler, W.H., Gradmann, S., Kimbell, G.S., McDermott, K., Petersen, U.K. (2016a). A review of the NE Atlantic conjugate margins based on seismic refraction data. *Geol. Soc. Lond. Spec. Pub.*, 447 [Online]. Available at: http://doi.org/10.1144/SP447.9.

Funck, T., Geissler, W.H., Kimbell, G.S., Gradmann, S., Erlendsson, Ö., McDermott, K., Petersen, U.K. (2016b). Moho and basement depth in the NE Atlantic Ocean based on seismic refraction data and receiver functions. *Geol. Soc. London Spec. Pub.*, 447 [Online]. Available at: http://doi.org/10.1144/SP447.1.

Gaina, C. and Jakob, J. (2018). Global Eocene tectonic unrest: Possible causes and effects around the North American plate. *Tectonophysics*, 760, 136–151.

Gaina, C., Nasuti, A., Kimbell, G.S., Blischke, A. (2017). Break-up and seafloor spreading domains in the NE Atlantic. In *The NE Atlantic Region: A Reappraisal of Crustal Structure, Tectonostratigraphy and Magmatic Evolution*, Peron-Pinvidic, G., Hopper, J.R., Stoker, M.S., Gaina, C., Doornenbal, J.C., Funck, T., Arting, U.E. (eds). Geological Society of London Special Publications, London.

Geoffroy, L. (2001). The structure of volcanic margins: Some problematics from the North-Atlantic/Labrador-Baffin system. *Marine Petrol. Geol.*, 18(4), 463–469.

Geoffroy, L. (2005). Volcanic passive margins. *C.R. Geosciences*, 337, 1395–1408.

Geoffroy, L., Callot, J.P., Scaillet, S., Skuce, A., Gélard, J.P., Ravilly, M., Angelier, J., Bonin, B., Cayet, C., Perrot, K., Lepvrier, C. (2001). Southeast Baffin volcanic margin and the North American-Greenland plate separation. *Tectonics*, 20(4), 566–584.

Geoffroy, L., Aubourg, C., Callot, P., Barrat, J.A. (2007). Mechanisms of crustal growth in large igneous provinces: The North Atlantic province as a case study. *Geol. Soc. Am.*, Spec. Pub., 430, 747–774.

Geoffroy, L., Burov, E.B., Werner, P. (2015). Volcanic passive margins: Another way to break up continents. *Scientific Rep.*, 5(14828) [Online]. Available at: http://doi.org/10.1038/srep14828.

Geoffroy, L., Guan, H., Gernigon, L., Foulger, G.R., Werner, P. (2020). The extent of continental material in oceans: C-Blocks and the Laxmi Basin example. *Geophys. J. Int.*, 222(3) [Online]. Available at: http://doi: 10.1093/gji/ggaa215.

Gernigon, L., Ringenbach, J.C., Planke, S., Le Gall, B., Jonquet-Kolst, Ø.H. (2003). Extension, crustal structure and magmatism at the outer Vøring Basin, Norwegian margin. *J. Geol. Soc. London*, 160(2), 197–208.

Gernigon, L., Blischke, A., Nasuti, A., Sand, M. (2015). Conjugate volcanic rifted margins, seafloor spreading, and microcontinent: Insights from new high-resolution aeromagnetic surveys in the Norway Basin. *Tectonics*, 34(5), 907–933.

Gernigon, L., Franke, D., Geoffroy, L., Schiffer, C., Foulger, G.R., Stoker, M. (2019). Crustal fragmentation, magmatism, and the diachronous opening of the Norwegian-Greenland Sea. *Earth Sci. Rev.* [Online]. Available at: http://doi.org/10.1016/j.earscirev.2019. 04.011.

Gilotti, J.A. and McClelland, W.C. (2008). The Greenland Caledonides: Evolution of the northeast margin of Laurentia. *Geol. Soc. Am.* [Online]. Available at: http://doi.org/10. 1130/MEM202.

Graham, D., Michael, P., Shea, T. (2016). Extreme incompatibility of helium during mantle melting: Evidence from undegassed mid-ocean ridge basalts. *Earth Planet. Sci. Lett.*, 454, 192–202.

Guan, H., Geoffroy, L., Gernigon, L., Chauvet, F., Grigné, C., Werner, P. (2019). Magmatic ocean-continent transitions. *Mar. Pet. Geol.*, 104, 438–450.

He, Y., Wen, L., Capdeville, Y., Zhao, L. (2015). Seismic evidence for an Iceland thermochemical plume in the Earth's lowermost mantle. *Earth Planet. Sci. Lett.*, 417, 19–27.

Hey, R., Martinez, F., Höskuldsson, A., Benediktsdóttir, A. (2010). Propagating rift model for the V-shaped ridges south of Iceland. *Geochem. Geophys. Geosyst.*, 11(Q03011) [Online]. Available at: http://doi.org/10.1029/2009GC002865.

Hjartarson, A., Erlendsson, Ö., Blischke, A. (2007). The Greenland-Iceland-Faroe Ridge complex. In *The NE Atlantic Region: A Reappraisal of Crustal Structure, Tectonostratigraphy and Magmatic Evolution*, Péron-Pinvidic, G., Hopper, J.R., Stoker, M., Gaina, C., Funck, T., Árting, U.E., Doornenbal, J.C. (eds). Geological Society of London, London.

Hole, M. and Natland, J. (2019). Magmatism in the North Atlantic Igneous Province: Mantle temperatures, rifting and geodynamics. *Earth Sci. Rev.* [Online]. Available at: doi.org/10. 1016/j.earscirev.2019.02.011.

Jóhanesson, H. and Sædmundsson, K. (1998). *Carte géologique de l'Islande à 1/500 000*, 2nd edition. Icelandic Institute of Natural History, Reykjavík.

Keen, C.E. and Boutilier, R.R. (1995). Lithosphere-asthenosphere interactions below rifts. In *Rifted Ocean-Continent Boundaries*, Banda, E., Torné, M., Talwani, M. (eds). Kluwer Academy, Norwell.

Kent, A.J.R., Stolper, E.M., Francis, D., Woodhead, J., Frei, R., Eiler, J. (2004). Mantle heterogeneity during the formation of the North Atlantic Igneous Province: Constraints from trace element and Sr-Nd-Os-O isotopesystematics of Baffin Island picrites. *Geochem. Geophys. Geosyst.*, 5(11) [Online]. Available at: http://doi.org/10.1029/ 2004GC000743.

Kerr, D.K. (1995). The melting processes and composition of the North Atlantic (Iceland) plume: Geochemical evidence from the Early Tertiary basalts. *J. Geol. Soc. London.*, 152, 975–978.

King, S.D. (2005). North Atlantic topographic and geoid anomalies: The result of a narrow ocean basin and cratonic roots? In Foulger, G.R., Natland, J.H., Presnall, J.H. and Anderson, D.L. (eds). *Plates, Plumes, and Paradigms*, Geological Society of America Special Paper, 388, doi:10.1130/2005.2388(37).

King, S.D. and Anderson, D. (1998). Edge-driven convection. *Earth Planet. Sci. Lett.*, 60, 289–296.

Klausen, M.B. and Larsen, H.C. (2002). East Greenland coast parallel dike swarm and its role in continental breakup. In *Volcanic Rifted Margins*, Menzies, M.A., Klemperer, S.L., Ebinger, C.J., Baker, J. (eds). The Geological Society of America, McLean, VA.

Koopmann, H., Schreckenberger, B., Franke, D., Becker, K., Schnabel, M. (2014). The late rifting phase and continental break-up of the southern South Atlantic: The mode and timing of volcanic rifting and formation of earliest oceanic crust. Geological Society, Special Publications, v.420, London [Online]. Available at: http//doi.org/10.1144/SP420.2.

Larsen, H.C. and Jakobsdóttir, S. (1988). Distribution, crustal properties and significance of seawards-dipping sub-basement reflectors off East Greenland. In *Early Tertiary Volcanism and the Opening of the Northeast Atlantic*, Morton, A.C., Parson, L.M. (eds). Geological Society of London, London.

Larsen, L.M., Heaman, L.M., Creaser, R.A., Duncan, R.A., Frei, R., Hutchison, M. (2009). Tectonomagmatic events during stretching and basin formation in the Labrador Sea and the Davis Strait: Evidence from age and composition of Mesozoic to Palaeogene dyke swarms in West Greenland. *J. Geol. Soc.*, 166, 999–1012.

Lawver, L.A. and Müller, R.D. (1994). Iceland hotspot track. *Geology*, 22(4), 311–314.

Lenoir, X., Féraud, G., Geoffroy, L. (2003). High-rate flexure of the East Greenland volcanic margin: Constraints from $^{40}Ar/^{39}Ar$ dating of basaltic dykes. *Earth Planet. Sci Lett.*, 214(3–4), 515–528.

Martinez, F., Hey, R., Höskuldsson, Á. (2019). Reykjanes Ridge evolution: Effects of plate kinematics, small-scale upper mantle convection and a regional mantle gradient. *Earth. Sci. Rev.* [Online]. Available at: https://doi.org/10.1016/j.earscirev.2019.102956.

Marty, B., Upton, B.G.J., Ellam, R.M. (1998). Helium isotopes in early Tertiary basalts, northeast Greenland: Evidence for 58 Ma plume activity in the north Atlantic-Iceland volcanic province. *Geology*, 26(5), 407–410.

Menke, W. and Sparks, D. (1995). Crustal accretion model for Iceland predicts "cold" crust. *Geophys. Res. Lett.*, 22, 1673–1676.

Mercur'ev, S.A., DeMets, C., Gurevich, N.I. (2009). Geodynamic evolution of crust accretion at the axis of the Reykjanes Ridge, Atlantic Ocean. *Geotectonics*, 43(3), 194–207.

Montagner, J.P. and Ritsema, J. (2001). Interactions between ridges and plumes. *Science*, 294(5546), 1472–1473.

Moreira, M. (2013). Noble gas constraints on the origin and evolution of Earth's volatiles. *Geochemical Perspectives, Vol. 2*. Institut physique du globe de Paris, Paris. doi: 10.7185/geochempersp.2.2.

Morency, C. and Doin, M.-P. (2004). Numerical simulations of the mantle lithosphere delamination. *J. Geophys. Res.*, 109(B03410) [Online]. Available at: http://doi.org/ 10.1029/ 2003JB002414.

Morency, C., Doin, M.-P., Dumoulin, C. (2002). Convective destabilization of a thickened continental lithosphere. *Earth Planet. Sci. Lett.*, 202(2), 303–320.

Oakey, G.N. and Chalmers, J.A. (2012). A new model for the Paleogene motion of Greenland relative to North America: Plate reconstructions of the Davis Strait and Nares Strait regions between Canada and Greenland. *J. Geophys. Res.*, 117(B10401) [Online]. Available at: http://doi.org/10.1029/ 2011JB008942.

Osmundsen, P.T. and Ebbing, J. (2008). Styles of extension offshore mid-Norway and implications for mechanisms of crustal thinning at passive margins. *Tectonics*, 27 [Online]. Available at: http://doi.org/10.1029/2007TC002242.

Parman, S.W., Kurz, M.D., Hart, S.R., Grove T.L. (2005). Helium solubility in olivine and implications for high ^{3}He/^{4}He in ocean island basalts. *Nature*, 437(7062), 1140–1143.

Petersen, K.D., Schiffer, C., Nagel, T. (2018). Formation and protracted T lower mantle upwelling induced by rifting and delamination. *Scientific Rep.*, 8(1) [Online]. Available at: http://doi.org/10.1038/s41598-018-34194-0.

Richards, M.A., Duncan, R.A., Courtillot, V. (1989). Flood basalts and hot spot tracks: Plume heads and tails. *Science*, 246, 103–107.

Ritsema, J., Jan van Heijst, H., Woodhouse, J.H. (1999). Complex shear wave velocity structure imaged beneath Africa and Iceland. *Science*, 286(5446), 1925–1928 [Online]. Available at: http://doi.org/10.1126/science.286.5446.1925.

Roest, W.R. and Srivastava, S.P. (1989). Seafloor spreading in the Labrador Sea: A new reconstruction. *Geology*, 17, 1000–1004.

Rogers, D.A., Marshall, J.E.A., Astin, T.R. (1989). Devonian and later movements on the Great Glen fault system, Scotland. *J. Geol. Soc. London*, 146(3), 369–372.

Saunders, A.D., Jones, S.M., Morgan, L.A., Pierce, K.L., Widdowson, M., Xu, Y.G. (2007). Regional uplift associated with continental large igneous provinces: The roles of mantle plumes and the lithosphere. *Chem. Geol.*, 241, 282–318.

Schiffer, C., Balling, N., Jacobsen, B.O., Stephenson, R.A., Nielsen, S.B. (2014). Caledonides seismological evidence for a fossil subduction zone in the East Greenland Caledonides. *Geology*, 42(4), 311–314.

Schiffer, C., Doré, A.G., Foulger, G.R., Franke, D., Geoffroy, L., Gernigon, L., Holdsworth, B., Kusznir, N., Lundin, E., McCaffrey, K. *et al.* (2019). Structural inheritance in the North Atlantic. *Earth Sci. Rev.* [Online]. Available at: http://doi.org/10.1016/j.earscirev. 2019.102975.

Sengör, A.M.C. and Burke, K. (1978). Relative timing of rifting and volcanism on Earth and its tectonic implications. *Geophys. Res. Lett.*, 5, 419–421 [Online]. Available at: http://doi.org/10.1029/GL005i006p00419.

Simon, K., Huismans, R.S., Beaumont, C. (2009). Dynamical modelling of lithospheric extension and small-scale convection: Implications for magmatism during the formation of volcanic rifted margins. *Geophys. J. Int.*, 176, 327–350.

Srivastava, S.P. and Roest, W.R. (1999). The extent of oceanic crust in the Labrador Sea. *Marine Petrol. Geol.*, 16, 65–84.

Suckro, S., Gohl, K., Funck, T., Heyde, I., Schreckenberger, B., Gerlings, J., Damm, V. (2013). The Davis Strait crust – A transform margin between two oceanic basins. *Geophys. J. Int.*, 193, 78–97.

Thompson, R.N. and Gibson, S.A. (1991). Subcontinental mantle plumes, hot spots and pre-existing thinspots. *J. Geol. Soc. London*, 148, 973–997.

Torsvik, T.H., Amundsen, H.E.F., Tronnes, R.G., Doubrovine, P.V., Gaina, C., Kusznir, N.J., Steinberger, B., Corfu, F., Ashwal, L.D., Griffin, W.L., Werner, S.C., Jamtveit, B. (2015). Continental crust beneath southeast Iceland. *PNAS*, 112(15), E1818–E1827.

Unternehr, P., Péron-Pinvidic, G., Manatschal, G., Sutra, E. (2019). Hyper-extended crust in the South Atlantic: In search of a model. *Petrol. Geosci.*, 16(3), 207–215.

Van der Hilst, R.D. and De Hoop, M.V. (2005). Banana-doughnut kernels and mantle tomography. *Geophys. J. Int.*, 163, 956–961.

Van Wijk, J.W. and Cloetingh, S.A. (2002). Basin migration caused by slow lithospheric extension. *Earth Planet. Sci. Lett.*, 198, 275–288.

Walker, G.P.L. (1959). Geology of the Reydarfjördur area, Eastern Iceland. *Quart. J. Geol. Soc. London*, 114, 367–393.

Watt, W.S. (1969). The coast-parallel dike swarm of southwest Greenland in relation to the opening of the Labrador Sea. *Can. J. Earth Sci.*, 6, 1320–1321.

White, N. and Lovell, B. (1997). Measuring the pulse of a plume with the sedimentary record. *Nature*, 387, 888–891.

White, R.S. and Mackenzie, D. (1989). Magmatism at rift zones: The generation of volcanic continental margins and flood basalts. *J. Geophys. Res.*, 94, 7685–7729.

White, R.S., Spence, G.D., Fowler, S.R., McKenzie, D.P., Westbrook, G.K., Bowen, A.N. (1987). Magmatism at rifted continental margins. *Nature*, 330, 439–444.

Wilkinson, C.M., Ganerød, M., Hendriks, B.H.K., Eide, E.A. (2016). Compilation and appraisal of geochronological data from the North-Atlantic Igneous Province (NAIP) . In *The NE Atlantic Region: A Reappraisal of Crustal Structure, Tectonostratigraphy and Magmatic Evolution*, Péron-Pinvidic, G., Hopper, J.R., Stoker, M., Gaina, C., Funck, T., Árting, U.E., Doornenbal, J.C. (eds). Geological Society of London, London [Online]. Available at: http://doi.org/10.1144/SP447.10.

Wilson, M., Neumann, E.R., Davies, G.R., Timmerman, M.T., Heeremans, M., Larsen, B.T. (2004). Permo-carboniferous magmatism and rifting in Europe. *Geol. Soc. London*, 223, 1–10.

Wolfe, C.J., Bjarnason, I.T., Van Decar, J.C., Solomon, S.C. (1997). Seismic structure of the Iceland mantle plume. *Nature*, 385, 245–247.

Yuan, X., Korenaga, J., Holbrook, W.S., Kelemen, P.B. (2020). Crustal structure of the Greenland–Iceland Ridge from joint refraction and reflection seismic tomography. *J. Geophys. Res.*, 12. Available at: https://doi.org/10.1029/2020JB019847.

Ziegler, P.A. (1989). *Evolution of the Arctic-North Atlantic and the Western Tethys.* AAPG, Tulsa, 164–196.

Ziegler, P.A. (1992). North Sea rift system. *Tectonophysics*, 208, 55–75 [Online]. Available at: http://doi.org/10.1016/0040-1951(92)90336-5.

Conclusion

Françoise BERGERAT and Laurent GEOFFROY with the collaboration of Brigitte VAN VLIET-LANOË and René MAURY

Iceland is not only a splendid country with its large ice caps, volcanoes and a rift that constitutes, in the middle of the island, the geological border between America and Europe. As we have just seen in this first volume, the genesis of the island is the result of abundant volcanism, mainly basaltic, linked to a thermal anomaly in the mantle with a debated origin, whose apex (hot spot) is located under the Vatnajökull glacier. The coupled functioning of this hot spot and the Mid-Atlantic Ridge has led to the formation of an island with a thickened crust, cut from north to south by a volcanic rift whose location is unstable over time. Moreover, the geological history of Iceland is fundamentally linked to that of the North Atlantic and, although young, it bears witness to a geodynamic evolution that dates back to the Paleozoic.

A long and complex history

The origin of Iceland and its nature are inseparable from the history of the breakup of Laurussia. The breakup of a supercontinent is a complex process involving the notions of structural and thermal inheritance of the continental lithosphere, the dynamic conditions at the limits of the supercontinent (peripheral subductions and associated stress fields) and finally the behavior of the isolated infralithospheric mantle under the supercontinent. This complex history is deeply marked within the evolution of the Atlantic domain. Both the suture zones in the lithospheric mantle (witnesses of

Iceland Within the Northern Atlantic 1,
coordinated by Brigitte VAN VLIET-LANOË. © ISTE Ltd 2021.

ancient plunging plates) and the thickened orogenic crusts of the Caledonian and Variscan axes have played a role in the location of the weakening zones of this supercontinent. Large-scale mantle melting marked just about every major stage of the separation of Laurussia: Carboniferous, Permian, Triassic, Jurassic and Paleogene. Scotland, with large swarms of dykes of Carboniferous, Permian and Paleocene ages, is an illustration of the probable covering of this region by kilometers of lava of distinct ages, now eroded. It is in this context that the Icelandic hot spot must also be interpreted.

Plume or no plume, that is the question

Tomographic studies show the presence under Iceland of a thermal anomaly in the upper mantle, with a diameter of approximately 200 km, which can be characterized at least to a depth of 400 km and whose apex is located under the Vatnajökull. Several authors have hypothesized that this anomaly is rooted within the lower mantle, or even at the core-mantle boundary, but its continuity at depths greater than 600 km is questionable. Two types of interpretation are proposed based on current knowledge. Many authors (notably petrologists and geochemists) consider it as a deep "classical" Hawaiian type plume: this opinion is consistent, especially, with the nature of the enriched components found in the lava sources and with the high $^{3}He/^{4}He$ ratios of the lava. Their arguments are presented in Chapter 1 of Volume 2. Others (notably geophysicists) propose tectonically dominant models in relation with small-scale convection processes. Indeed, the distribution of lava of Paleocene-Eocene age is closely linked to the reactivation of shear zones of lithospheric significance (such as the Great Glen Fault in Scotland), or to the lithospheric extension located at the edges of the cold and thick Greenland lithosphere. Mantle melting thus seems to be correlated above all with the lithospheric thinning by extension caused by regional tectonic stresses and with the structural and thermal inheritance of the continental lithosphere. The existence of an asthenosphere slightly warmer than "normal" may have various origins, such as convection at the edge of the craton, although a convincing explanation must be given, notably, for the high helium isotope ratios.

An oceanic rift with special characteristics

One of the particular characteristics of the active Icelandic Rift is its youth (<9 My for the oldest segment), which implies the existence of older rifts. The location of these rifts and the exact period of their functioning are still far from being known with certainty. Characterized from the age of the lavas, paleomagnetic data and structural observations, they are dependent on these data and thus on their uncertainties. The last paleo-rift in the north of the island is classically located in the Húnaflói-Skagi peninsula. It would have been active until 1–2 My and would have functioned simultaneously with the North Volcanic Zone for 4–6 My. However, new dating, based on the age of the dykes, are in favor of a location of the last rift further east, in the Skagafjörður, active until about 3 My and contemporary of the current rift for 5–5.5 My. The existence of an older paleo-rift, off the northwestern peninsula, is not excluded, although difficult to document. In the south, the rift jump is in progress, with the western and East Volcanic Zones functioning together for 2–3 My. A marked unconformity further west argues for the existence of an earlier paleo-rift, having operated between 15 and 6–7 My, which could be located in the Snæfells Peninsula.

Another specific feature of the active rift in Iceland is its width – which can reach almost 200 km when considering the periods when two parallel rift segments are functioning simultaneously – considerably greater than that of the submerged slow ridges. Rollover systems in connection with magma chambers have been suggested to explain this difference, but glacial loading can also lead to a widening of the active rift zone.

Each of the three branches of the active rift is associated with central volcanoes and consists of fissure swarms arranged *en échelon*. Although the deformation processes are different, fissures and normal faults contribute jointly to the extension, leading for the first ones to an increase in volume due to magmatic feeds (dykes) and for the normal faults to an elongation of the deformed zones. These processes result in asymmetrical grabens and tilted block systems. The partitioning of the deformation between these two modes of deformation evolves with depth, as shown by the comparison between the structures of the active rift and the – deeper – structures visible in the eroded older lava flows.

Seismic activity in the rift is not associated with the tectonic process of rifting itself, but with the functioning of magmatic systems and in particular with the propagation of dykes. On the other hand, in the fault zones connecting the Icelandic Rift to the Kolbeinsey Ridge to the north and to the Reykjanes Ridge to the south, seismicity reflects the transform process in all its complexity and diversity from its immature to mature stages. The first one is illustrated by the South Iceland Seismic Zone, where the E-W sinistral transform play is accommodated by *en échelon* dextral N-S faults. The second one is represented by the Tjörnes Fracture Zone where the seismic lineaments, ESE-WNW, are more or less parallel to the transform direction. Within the latter, however, faults with different degrees of development coexist.

The comparison between the stresses determined by the focal mechanisms of earthquakes and those revealed by the analysis of older faults shows that the partitioning of the deformation can exist on the scale of a seismic crisis as well as on that of a few million years.

This question of the time scale considered should not be ignored in Iceland. It should be kept in mind that seismological data or GPS data only reflect a "snapshot" of motions and deformation, and do not necessarily represent a general process that has been going on for – or will maintain for – a few million years.

A still mysterious island

Iceland remains a mysterious island in many ways. Our knowledge of the deep structure of the Icelandic crust is limited to geophysical data, such as seismic wave velocity variations and gravimetric and magnetotelluric inversions. These data are not unambiguously interpretable and do not allow characterization of the nature and composition of Iceland's lower crust, the thickness of which is considerable and highly variable laterally. The accretionary process of the Icelandic mafic crust led to the formation of SDRs in the upper crust of the island. However, no known slow-spreading oceanic ridge produced basaltic lava prisms of multi-kilometer thickness dipping toward the oceanic rift axis. The SDRs in Iceland have been very little studied and the structure of the upper crust of the island is particularly poorly constrained. The intimate structure of these SDRs can be observed when they are sufficiently eroded. They are associated with kilometric offset normal faults dipping not toward the active rift, but toward the distal volcanic margins of Greenland and Faroe Islands. The crustal structure of Iceland, like that of the Greenland–Iceland–Faroe Islands ridge as a whole, presents surprising analogies with that of the adjacent passive volcanic (continental) margins. Arguments in favor of the presence of continental material in the sub-basaltic crust of the Greenland–Iceland–Faroe Ridge are growing. Many aspects of the geodynamic evolution of this complex region of the globe, since the Paleogene, remain unexplained and their causes have yet to be discovered.

The geodynamic evolution of Iceland in the vast North Atlantic Ocean and its own tectonic history are, as we have seen throughout these pages, intimately linked to an abundant volcanism, mainly basaltic (production of 80,000 km^3 of lava per My for the last 16 My). The complex geochemical signature of Icelandic basalts is consistent with their derivation from mixtures (in variable proportions) of a depleted sub-oceanic upper mantle similar to the source of MORB, and of enriched mantle components typical of intra-plate oceanic islands. Another aspect of the external geodynamics of the island, which has been only briefly skimmed in these pages, is the presence – and role – of its ice cap. The Icelandic landscape is largely dependent on episodes of glaciation and deglaciation, and on the huge volumes of lava emitted during the Middle Miocene and Pliocene, further abraded by glaciers, the erosion caused by glacially dependent episodes being much significant than volcanic contributions. Finally, because of its position in the middle of the Atlantic and of its ice cover, Iceland is a recorder and witness of major climate changes. This Iceland of volcanoes and glaciers is presented in Volume 2.

References

Key books

Aber, J.S., Croot, D.F., Fenton, M.M. (1989). *Glacitectonic Landforms and Structures*. Kluwer, Dordrecht.

Allen, P.A. (1997). *Earth Surface Processes*. Blackwell, Oxford.

Arnalds, O. (2015). *The Soils of Iceland*. Springer, Dordrecht.

Bardintzeff, J.-M. (2016). *Volcanologie*, 5th edition. Dunod, Paris.

Benn, D.I. and Evans, D.J. (2011). *Glaciers & Glaciations*. Hodder Arnold Publication, London.

Björnsson, H. (2019). *The Glaciers of Iceland*. Atlantis Press, Springer, Amsterdam. Available at: http///wwwspringer.com/series/15358.

Broecker, W. (2010). *The Great Ocean Conveyor: Discovering the Trigger for Abrupt Climate Change*. Princeton University Press, Princeton.

Cazenave, A. and Feigl, K. (1994) *Formes et mouvements de la terre*. Belin, Paris.

Chazot, G., Lenat, J-F., Maury, R., Agranier, A., Roche, O. (2017). *Volcanologie*. De Boeck Superieur, Louvain-La-Neuve.

Cuffey, K.M. and Paterson, W.S. (2010). *Physics of Glaciers*, 4th edition. Elsevier, Oxford.

Detay, M. (2017). *Traité de volcanologie physique*. Lavoisier, Paris.

Detay, M. and Detay, A.-M. (2013). *Volcans. Du feu et de l'eau*. Belin, Paris.

Dickson, R., Denk, T., Grímsson, F., Zetter, R., Símonarson, L.A. (2011). *Late Cainozoic Floras of Iceland–15 Million Years of Vegetation and Climate History in the Northern North Atlantic*. Springer Verlag, Dordrecht.

Duplessy, J.C. and Morel, P. (1990). *Gros temps sur la planète*. Odile Jacob, Paris.

Duplessy, J.C. and Ramstein, G. (2013). *Paléoclimatologie. Tome 1 : trouver, dater et interpréter les indices : enquête sur les climats anciens*. EDP Sciences, Paris.

Edwards, B.R., Guðmundsson, M.T., Russell, J.K. (2015). Glaciovolcanism. In *Encyclopedia of Volcanoes*, 2nd edition, Sigurdsson, H., Houghton, B., Rymer, H., Stix, J., McNutt, S. (eds). Elsevier, Oxford.

Ehlers, J., Gibbard, P., Hughes, P.D. (2011). *Quaternary Glaciations. Extent and Chronology: A Closer Look*. Elsevier, Oxford.

Einarsson, Þ. (1994). *Geology of Iceland. Rocks and Landscape*. Mál og menning, Reykjavík.

Francis, P. (1993). *Volcanoes: A Planetary Perspective*. Clarendon Press, Oxford.

Gaudru, H. and Chazot, G. (2018). *La belle histoire des volcans*. De Boeck Supérieur, Louvain-la-Neuve.

Gill, R. (2010). *Igneous Rocks and Processes–A Practical Guide*. Wiley-Blackwell, Oxford.

Guðmundsson, Á. (2011). *Rock Fractures in Geological Processes*. Cambridge University Press, Cambridge.

Guðmundsson, Á. (2017). *The Glorious Geology of Iceland's Golden Circle*. Springer, Cham.

Henriksen, N. (2005). *Geological History of Greenland. Four Billion Years of Earth Evolution*. Geological Survey of Denmark and Greenland, Copenhagen.

Hjálmarsson, J. (1993). *History of Iceland*. Iceland Review, Reykjavík.

Juteau, T. and Maury, R.C. (2012). *La croûte océanique. Pétrologie et dynamique endogènes*. Vuibert, Paris.

Kraft, M. and Kraft, K. (1974). *Guide des volcans d'Europe : généralités, France, Islande, Italie, Grèce, Allemagne*. Delachaux et Niestlé, Neuchâtel.

Lagabrielle, Y., Maury, R., Renard, M. (2017). *Mémo visuel de géologie*, 2nd edition. Dunod, Paris.

Meyen, S.V. (1987). *Fundamentals of Palaeobotany*. Chapman & Hall, London.

Oppenheimer, C. (2011). *Eruptions that Shook the World*. Cambridge University Press, Cambridge.

Ruddiman, E.W. (ed.) (1999). *Tectonic Uplift and Climate Change*. Plenum Press, New York.

Ruddiman, E.W. (2008). *Earth's Climate: Past and Future*. Freeman & Co., New York.

Schminke, H.-U. (2004). *Volcanism*. Springer, Berlin.

Siebert, L., Simkin, T., Kimberly, P. (2011). *Volcanoes of the World*, 3rd edition. University of California Press, Berkeley.

Sigmundsson, F. (2006). *Iceland Geodynamics. Crustal Deformation and Divergent Plate Tectonics*. Springer, Chichester.

Sigurdsson, H. (ed.). (2000). *Encyclopedia of Volcanoes*. Academic Press, London.

Spencer, A.M., Embry, A.F., Gautier, D.L., Stoupakova, A.V., Sørensen, K. (2011). *Arctic Petroleum Geology*. Geological Society of London, London [Online]. Available at: http://doi.org/10.1144/M35.

Thomas, D.N. and Dieckmann, G.S. (eds). (2010). *Sea Ice*. Wiley, Chichester.

Van Vliet-Lanoë, B. (2013). *Cryosphère : soixante millions d'années d'évolution de notre planète*. Vuibert, Paris.

Van Vliet-Lanoë, B. (2014). *Environnements froids*. Vuibert, Paris.

Van Vliet-Lanoë, B. (ed.) (2021). *Iceland Within the Northern Atlantic 2*. ISTE Ltd, London, and John Wiley & Sons, New York.

Zimbelman, J.R. and Gregg, T.K.P. (2000). *Environmental Effects on Volcanic Eruptions: From Deep Oceans to Deep Space*. Kluwer Academic/Plenum Publishers, New York.

Articles, documents and key theses

Acocella, V., Guðmundsson, Á., Funiciello, R. (2000). Interaction and linkage of extension fractures and normal faults: Examples from the rift zone of Iceland. *J. Struct. Geol.*, 22, 1233–1246.

Agranier, A., Maury, R.C., Geoffroy, L., Chauvet, F., Le Gall, B., Aviana, A. (2019). Volcanic record of continental thinning in Baffin Bay margins: Insights from Svartenhuk Halvø Peninsula basalts, West Greenland. *Lithos*, 334–335, 117–140.

Albino, F., Pinel, V., Sigmundsson, F. (2010). Influence of surface load variations on eruption likelihood: Application to two Icelandic subglacial volcanoes, Grimsvötn and Katla. *Geophys. J. Int.*, 181, 1510–1524.

Anderson, D.L. (2001). Top-down tectonics? *Science*, 293, 2016–2018.

Andrews, J.T., Hardarddóttir, J., Helgadóttir, G., Jennings, A.E., Geirsdóttir, A., Sveinbjornsdóttir, A.E., Schoolfield, S., Kristjansdóttir, G.B., Smith, L.M., Thors, K., Syvitski, J.P.M. (2000). The N and W Iceland shelf: Insights into Last Glacial Maximum ice extent and deglaciation based on acoustic stratigraphy and basal radiocarbon AMS dates. *Quat. Sci. Rev.*, 19, 619–631.

Andrews, J.T., Darby, D.A., Eberl, D.D., Jennings, A.E., Moros, M., Ogilvie, A. (2009). A robust multi-site Holocene history of drift ice off Northern Iceland: Implications for North Atlantic climate. *Holocene*, 19, 71–78.

Angelier, J., Bergerat, F., Dauteuil, O., Villemin, T. (1997). Effective tension-shear relationships in extensional fissure swarms, axial rift of northeastern Iceland: Morphological evidences. *J. Struct. Geol.*, 19, 673–685.

Angelier, J., Bergerat, F., Stefánsson, R., Bellou, M. (2008). Seismotectonics of a newly formed transform zone near a hot spot: Earthquake mechanisms and regional stress in the South Iceland Seismic Zone. *Tectonophysics*, 447, 95–116.

Arnórsson, S. (1995a). Geothermal systems in Iceland: Structure and conceptual models I. High temperature areas. *Geothermics*, 24(5–6), 561–602.

Arnórsson, S. (1995b). Geothermal systems in Iceland: Structure and conceptual models II. Low temperature areas. *Geothermics*, 24(5–6), 603–629.

Arnórsson, S., Axelsson, G., Sæmundsson, K. (2008). Geothermal systems in Iceland. *Jökull*, 58, 269–302.

Bellou, M. (2006). Analyse sismotectonique de la zone sismique sud-islandaise. PhD Thesis, Université Pierre et Marie Curie, Paris et Villefranche-sur-Mer.

Berger, A. (2004). Crises volcano-tectoniques et divergence de plaques en Islande : mesure par GPS et modélisation numérique. PhD Thesis, Université de Savoie, Chambéry.

Bergerat, F. and Angelier, J. (2008). Immature and mature transform zones near a hot spot: The South Iceland Seismic Zone and the Tjörnes Fracture Zone (Iceland). *Tectonophysics*, 447, 142–154.

Bergerat, F., Angelier, J., Homberg, C. (2000). Tectonic analysis of the Husavik-Flatey fault (Northern Iceland) and mechanisms of an oceanic transform zone, the Tjörnes Fracture Zone. *Tectonics*, 19, 1161–1177.

Bergerat, F., Angelier, J., Guðmundsson, Á., Torfason, H. (2003). Push-ups, fracture patterns, and paleoseismology of the Leirubakki Fault, South Iceland. *J. Struct. Geol.*, 25, 591–609.

Bergerat, F., Homberg, C., Angelier, J., Bellou, M. (2011). Surface traces of the Minnivellir, Réttarnes and Tjörvafit seismic faults in the South Iceland Seismic Zone: Segmentation, lengths and magnitude of related earthquakes. *Tectonophysics*, 498, 11–26.

Biessy, G., Dauteuil, O., Van Vliet-Lanoë, B., Wayolle, A. (2008). Fast and partitioned postglacial rebound of south-western Iceland. *Tectonics*, 27 [Online]. Available at: http://doi.org/10.1029/2007TC002177.

Boers, N., Ghil, M., Rousseau, D.D. (2018). Ocean circulation, ice shelf, and sea ice interactions explain Dansgaard-Oeschger cycles. *Proc. Nat. Acad. Sci.*, 115(47), E11005–E11014 [Online]. Available at: http://doi.org/10.1073/pnas.1802573115.

Bourgeois, O. (1998). Processus d'extension lithosphérique en Islande. Interactions avec les calottes glaciaires quaternaires. PhD Thesis, Université de Rennes 1, Rennes.

Bourgeois, O., Dauteuil, O., Van Vliet-Lanoë, B. (1998). Subglacial volcanism in Iceland: Tectonic implications. *Earth Planet. Sci. Lett.*, 164(1–2), 165–178.

Bourgeois, O., Dauteuil, O., Hallot, E. (2005). Rifting above a mantle plume: Structure and development of the Iceland Plateau. *Geodin. Acta*, 18(1), 1–22.

Cabedo-Sanz, P., Belt, S.T., Jennings, A.E., Andrews, J.T., Geirsdóttir, A. (2016). Variability in drift ice export from the Arctic Ocean to the North Icelandic Shelf over the last 8,000 years: A multi-proxy evaluation. *Quat. Sci. Rev.*, 146, 99–115.

Callot, J.P. (2002). Origine, structure et développement des marges volcaniques : l'exemple du Groenland : interactions manteau-lithosphère en contexte de panache. PhD Thesis, Université Pierre et Marie Curie, Paris.

Campbell, I.H. (2005). Large igneous provinces and the mantle plume hypothesis. *Elements*, 1, 265–269.

Campbell, I.H. and Kerr, A.C. (2007). The Great Plume Debate: Testing the plume theory. *Chem. Geol.*, 241(3–4), 149–152.

Chalmers, J.A. and Laursen, K.H. (1995). Labrador Sea: The extent of continental and oceanic crust and the timing of the onset of seafloor spreading. *Mar. Petrol. Geol.*, 12, 205–217.

Chauvet, F., Geoffroy, L., Guillou, H., Maury, R.C., Le Gall, B., Agranier, A., Aviana, A. (2019). Eocene continental breakup in Baffin Bay. *Tectonophysics*, 757, 170–186.

Class, C. and Goldstein, S.L. (2005). Evolution of helium isotopes in the Earth's mantle. *Nature*, 436, 1107–1112.

Coffin, M.F. and Eldholm, O. (1994). Large igneous provinces: Crustal structure, dimensions, and external consequences. *Rev. Geophys.*, 32, 1–36.

Courtillot, V., Davaille, A., Besse, J., Stock, J. (2003). Three distinct types of hotspots in the Earth's mantle. *Earth Planet Sci. Lett.*, 205, 295–308.

Darbyshire, F.A., White, R.S., Priestley, K.F. (2000). Structure of the crust and uppermost mantle of Iceland from a combined seismic and gravity study. *Earth Planet. Sci. Lett.*, 181(3), 409–428.

Dauteuil, O., Angelier, J., Bergerat, F., Verrier, S., Villemin, T. (2001). Deformation partitioning inside a fissure swarm of the northern Icelandic rift. *J. Struct. Geol.*, 23, 1359–1372.

Davies, S.M., Abbott, P.M., Meara, R.H., Pearce, N.J.G., Austin, W.E.N., Chapman, M.R., Svensson, A., Bigler, M., Rasmussen, T.L., Rasmussen, S.O., Farmer, E.J. (2014). A North Atlantic tephrostratigraphical framework for 130-60kab2k: New tephra discoveries, marine-based correlations, and future challenges. *Quat. Sci. Rev.*, 106, 101–121.

DeMets, C., Gordon, R.G., Argus, D.F. (2010). Geologically current plate motions. *Geophys. J. Int.*, 181, 1–80.

DePaolo, D.J. and Manga, M. (2003). Deep origin of hotspots–The mantle plume model. *Science*, 300, 920–921.

Detay, M. (2010). Éruption de l'Eyjafjöll, un volcan qui a du panache. *Pour la science*, 392, 70–76.

Detay, M. and Hróarsson, B. (2011). Les tunnels de lave. *Pour la science*, 399, 2–7.

Dossier pour la science. (2010). La terre à cœur ouvert. *Dossier pour la science*, 67(April–June), 1–119.

Døssing, A., Japsen, P., Watts, A.B., Nielsen, T., Jokat, W., Thybo, H., Dahl-Jensen, T. (2016). Miocene uplift of the NE Greenland margin linked to plate tectonics: Seismic evidence from the Greenland Fracture Zone, NE Atlantic: Margin Uplift and Plate Tectonics. *Tectonics*, 35(2), 257–282.

Doubre, C. (2004). Structure et mécanismes des segments de rift volcano-tectoniques : études de rifts anciens (Écosse, Islande) et d'un rift actif (Asal-Ghoubbet). PhD Thesis, Université du Maine, Le Mans.

Dubois, L. (2006). Étude mécanique de la crise sismique sud-islandaise de juin 2000 par modélisation numérique tridimensionnelle : effets rhéologiques et géométriques. PhD Thesis, Université Paul Sabatier, Toulouse.

Dyment, J., Lin, J., Baker, E. (2007). Ridge-hotspot interactions. What mid-ocean ridges tell us about deep earth processes. *Oceanography*, 20(1), 102–115.

Einarsson, P. (1991). Earthquakes and present-day tectonism in Iceland. *Tectonophysics*, 189, 261–279.

Einarsson, P. (2010). Mapping of Holocene surface ruptures in the South Iceland Seismic Zone. *Jökull*, 60, 117–134.

Eksinchol, I., Rudge, J.F., Maclennan, J. (2019). Rate of melt ascent beneath Iceland from the magmatic response to deglaciation. *Geochem. Geophys. Geosys.*, 20(6), 2585–2605.

Eldholm, O. and Grue, K. (1994). North Atlantic volcanic margins: Dimensions and production rates. *J. Geophys. Res.*, 99(B2), 2955–2968.

Eldrett, J.S., Harding, I.C., Wilson, P.A., Butler, E., Roberts, A.P. (2007). Continental ice in Greenland during the Eocene and Oligocene. *Nature*, 446, 176–179.

Fitton, J.G., Saunders, A.D., Norry, M.J., Hardason, B.S., Taylor, R.N. (1997). Thermal and chemical structure of the Iceland plume. *Earth Planet. Sci. Lett.*, 153, 197–208.

Foulger, G.R. and Anderson, D.L. (2005). A cool model for the Iceland hotspot. *J. Volcan. Geotherm. Res.*, 141, 1–22.

Foulger, G.R., Du, Z., Julian, B.R. (2003). Icelandic-type crust. *Geophys. J. Int.*, 155, 567–590.

Garcia, S. (2003). Implications d'un saut de rift et du fonctionnement d'une zone transformante sur les déformations du nord de l'Islande. Approches structurale, sismotectonique et radiochronologique. PhD Thesis, Université Pierre et Marie Curie, Paris.

Garcia, S., Angelier, J., Bergerat, F., Homberg, C., Dauteuil, O. (2008). Influence of rift jump and excess loading on the structural evolution of Northern Iceland. *Tectonics*, 27, TC1006 [Online]. Available at: http://doi.org/10.1029/2006TC002029.

Geoffroy., L. (2005). Volcanic passive margins. *C.R. Geosciences*, 1, 337, 1395–1408.

Guðmundsdóttir, E.R., Larsen, G., Björck, S., Ingólfsson, Ó., Striberger, J. (2016). A new high-resolution Holocene tephra stratigraphy in eastern Iceland: Improving the Icelandic and North Atlantic tephrochronology. *Quat. Sci. Rev.*, 150, 234–249.

Guðmundsson, Á. (2000). Dynamics of volcanic systems in Iceland. Example of tectonism and volcanism at juxtaposed hot spot and mid-ocean ridge systems. *Ann. Rev. Earth Planet.*, 28, 107–140.

Guðmundsson, Á. (2007). Infrastructure and evolution of ocean-ridge discontinuities in Iceland. *J. Geodyn.*, 43, 6–29.

Guðmundsson, M.T., Jónsdóttir, K., Hooper, A., Holohan, E., Halldórsson, S., Ófeigsson, B., Cesca, S., Vogfjörð, K.S., Sigmundsson, F., Högnadóttir, T., Einarsson, P., Sigmarsson, O., Jarosch, A.H., Jonasson, K., Magnusson, E., Hreinsdóttir, S., Bagnardi, M., Parks, M.M., Hjörleifsdóttir, V., Palsson, F., Walter, T.R., Schöpfer, M.P.J., Heimann, S., Reynolds, H.I., Dumont, S., Bali, E., Guðfinnsson, G.H., Dahm, T., Roberts, M.J., Hensch, M., Belart, J.M.C., Spaans, K., Jakobbsson, S., Guðmundsson, G.B., Rüshuus, M.S., Pedersen, G.M.B., van Boeckel, T., Oddsson, B., Pfeffer, M.A., Barsotti, S., Bergsson, B., Donovan, A., Borton, M.R., Aiuppa, A. (2016). Gradual caldera collapse at Bárdarbunga volcano, Iceland, regulated by lateral magma outflow. Science, 353(6296) [Online]. Available at : http://doi.org/10.1126/science.aaf8988.

Guillou, H., Scao, V., Nomade, S., Van Vliet-Lanoë, B., Liorzou, C., Guðmundsson, Á. (2019). $^{40}Ar/^{39}Ar$ dating of the Thorsmork ignimbrite and Icelandic sub-glacial rhyolites. *Quat. Sci. Rev.*, 209, 52–62.

Hanan, B.B. and Schilling, J.G. (1997). The dynamic evolution of the Iceland mantle plume: The lead isotope perspective. *Earth Planet. Sci. Lett.*, 151, 43–60.

Hansen, J., Jerram, D.A., Caffrey, K.M., Passey, S.R. (2009). The onset of the North Atlantic Igneous Province in a rifting perspective. *Geol. Mag.*, 146(3), 309–325.

Henriot, O. (2003). La déformation actuelle au nord de l'Islande, à la jonction entre un rift et une transformante : mesure par InSAR et modélisation d'un système volcano-tectonique actif. PhD Thesis, Université de Savoie, Chambéry.

Hjartarson, A., Erlendsson, Ö., Blischke, A. (2007). The Greenland-Iceland-Faroe Ridge complex. In *The NE Atlantic Region: A Reappraisal of Crustal Structure, Tectonostratigraphy and Magmatic Evolution*, Péron-Pinvidic, G., Hopper, J.R., Stoker, M., Gaina, C., Funck, T., Árting, U.E., Doornenbal, J.C. (eds). Geological Society of London, London.

Huybers, P.J. and Langmuir, C. (2009). Feedback between deglaciation, volcanism, and atmospheric CO_2. *Earth Planet. Sc. Lett.*, 286(3–4), 479–491.

Jakobsson, S.P., Jonasson, K., Sigurdsson, I.A. (2008). The three igneous rock series of Iceland. *Jökull*, 58, 117–138.

Jakobsson, M., Andreassen, K., Bjarnadottir, L.R., Dove, D., Dowdeswell, J.A., England, J.H., Funder, S., Hogan, K., Ingolfsson, O., Jennings, A., Larsen, N.K., Kirchner, N., Landvik, J.Y., Mayer, L., Mikkelsen, N., Möller, P., Niessen, F., Nilsson, J., O'Regan, M., Polyak, L., Norgaard-Pedersen, N., Stein, R. (2014). Arctic Ocean glacial history. *Quat. Sci. Rev.*, 92, 40–67.

Jansen, E., Fronval, T., Frank, R., Channell, J.E. (2000). Pliocene-Pleistocene ice rafting history and cyclicity in the Nordic Seas during the last 3.5 Myr. *Paleoceanography*, 15(6), 709–721.

Japsen, P., Green, P.F., Bonow, J.M., Nielsen, T.F.D., Chalmers, J.A. (2014). From volcanic plains to glaciated peaks: Burial and exhumation history of southern East Greenland after opening of the NE Atlantic. *Global Planet. Ch.*, 116, 91–114.

Jóhanesson, H. and Sædmundsson, K. (1998). Carte géologique de l'Islande à 1/500000, 2nd édition. Document, Icelandic Institute of Natural History, Reykjavík.

Jull, M. and McKenzie, D. (1996). The effect of deglaciation on mantle melting beneath Iceland. *J. Geophys. Res.*, 101, 21815–21828.

Kleiven, H.F., Jansen, E., Fronval, T.M., Smith, I. (2002). Intensification of northern hemisphere glaciations in the circum Atlantic region (3.5-2.4 Ma): Ice-rafted detritus evidence. *Palaeogeogr. Palaeoclim., Palaeoecol.*, 1(84), 213–223.

Kokfelt, T.F., Hoernle, K., Hauff, F., Fiebig, J., Werner, R., Garbe-Schonberg, D. (2006). Combined trace element and Pb, Nd, Sr, O isotope evidence for recycled oceanic crust (upper and lower) in the Iceland mantle plume. *J. Petrol.*, 47, 1705–1749.

Lacasse, C. and Garbe-Schönberg, C.D. (2001). Explosive silicic volcanism in Iceland and the Jan Mayen area during the last 6 Ma: Sources and timing of major eruptions. *J. Volcan. Geoth. Res.*, 107, 113–147.

Lang, N. and Wolff, E.W. (2011). Interglacial and glacial variability from the last 800 ka in marine, ice and terrestrial archives. *Clim. Past*, 7(3), 61–380.

Lawver., L.A. and Müller, R.D. (1994). Iceland hotspot track. *Geology*, 22(4), 311–314.

Licciardi, J.M., Kurz, M.D., Curtice, J.M. (2007). Glacial and volcanic history of Icelandic table mountains from cosmogenic ^{3}He exposure ages. *Quat. Sci. Rev.*, 26, 1529–1546.

Lisiecki, L.E. and Raymo, M.E. (2005). A Pliocene-Pleistocene stack of 57 globally distributed benthic δ^{18}O records. *Paleoceanography*, 20 [Online]. Available at: http://doi.org/10.1029/2004PA001071.

Maclennan, J., Jull, M., McKenzie, D., Slater, D., Grönvold, K. (2002). The link between volcanism and deglaciation in Iceland. *Geoch. Geophys. Geosys.*, 3, 1062–1087.

Martin, E. and Sigmarsson, O. (2010). Thirteen million years of silicic magma production in Iceland: Links between petrogenesis and tectonic setting. *Lithos*, 116, 129–144.

McDougall, I., Kristjansson, L., Sæmundsson, K. (1984). Magnetostratigraphy and geochronology of NW Iceland. *J. Geophys. Res.*, 89(B8), 7029–7060.

Meyer, R., Van Wijk, J., Gernigon, L. (2007). The North Atlantic Igneous Province: A review of models for its formation. *Geol. Soc. Amer.*, 430, 525–552.

Montelli, R., Nolet, G., Dahlen, A., Masters, G.A. (2006). Catalogue of deep mantle plumes: New results from finite-frequency tomography. *Geochem. Geophys. Geosys.*, 7 [Online]. Available at: http://doi.org/10.1029/2006GC001248.

Peate, D.W., Breddam, K., Baker, J.A., Kurtz, M., Barker, A.K., Prestvik, T., Grassineau, N., Skovgaard, A.C. (2010). Compositional characteristics and spatial distribution of enriched Icelandic mantle components. *J. Petrol.*, 51, 1447–1475.

Plateaux, R. (2012). Architecture et mécanismes du rift islandais dans la région du Vatnajökull. PhD Thesis, Université de Nice-Sophia-Antipolis, Nice.

Rasmussen, T.L., Thomsen, E., Moros, M. (2016). North Atlantic warming during Dansgaard-Oeschger events synchronous with Antarctic warming and out-of-phase with Greenland climate. *Nature Sci. Rep.*, 6(20535) [Online]. Available at: http://doi.org/10.1038/srep20535.

Rögnvaldsson, S.T., Guðmundsson, Á., Slunga, R. (1998). Seismotectonic analysis of the Tjörnes Fracture Zone, an active transform fault in north Iceland. *J. Geophy. Res.*, 103(B12), 30117–30129.

Sæmundsson, K. (1979). Outline of the geology of Iceland. *Jökull*, 29, 7–28.

Saunders, A.D., Jones, S.M., Morgan, L.A., Pierce, K.L., Widdowson, M., Xu, Y.G. (2007). Regional uplift associated with continental large igneous provinces: The roles of mantle plumes and the lithosphere. *Chem. Geol.*, 241, 282–318.

Schiffer, C., Doré, A.G., Foulger, G.R., Franke, D., Geoffroy, L., Gernigon, L., Holdsworth, B., Kuznir, N.J., Lundin, E., McCaffrey, K., Peace, A.R., Petersen, K.D., Phillips, T.B., Stephenson, R., Stoker, M.S., Wellford, J.K. (2019). Structural inheritance in the North Atlantic. *Earth Sci. Rev.* [Online]. Available at: http://doi.org/10.1016/j.earscirev.2019.102975.

Sigmarsson, O. and Steinthorsson, S. (2007). Origin of Icelandic basalts: A review of their petrology and geochemistry. *J. Geodyn.*, 43, 87–100.

Sigmarsson, O., Maclennan, J., Carpentier, M. (2008). Geochemistry of igneous rocks in Iceland: A review. *Jökull*, 58, 139–160.

Sigmundsson, F., Einarsson, P., Bilham, R., Sturkell, E. (1995). Rift-transform kinematics in South Iceland: Deformation from Global Positioning System measurements, 1986–1992. *J. Geophys. Res.*, 100, 6235–6248.

Sigmundsson, F., Einarsson, P., Hjartardóttir, Á.R., Drouin, V., Jónsdóttir, K., Árnadóttir, Þ., Geirsson, H., Hreinsdóttir, S., Li, S., Ófeigsson, B.G. (2018). Geodynamics of Iceland and the signatures of plate spreading. *J. Volcan. Geotherm. Res.* [Online]. Available at: http://doi.org/10.1016/j.jvolge ores.2018. 08.014.

Stefánsson, R., Guðmundsson, G.B., Halldorsson, P. (2008). Tjörnes fracture zone. New and old seismic evidences for the link between the North Iceland rift zone and the Mid-Atlantic ridge. *Tectonophysics*, 447, 117–126.

Tarduno, J. (2010). La mobilité des points chauds. *Dossier pour la science*, 67, 56–61.

Thordarson, T. and Larsen, G. (2007). Volcanism in Iceland in historical times: Volcano types, eruption styles and eruptive history. *J. Geodyn.*, 43, 118–152.

Thordarson, T. and Self, S. (2003). Atmospheric and environmental effects of the 1783–1784 Laki eruption: A review and reassessment. *J. Geophys. Res.*, 108(D1), 4011.

Torsvik, T.H., Amundsen, H.E.F., Tronnes, R.G., Doubrovine, P.V., Gaina, C., Kuznir, N.J. Steinberger, B., Corfu, F., Ashwal, L.D., Griffin, W.L., Werner, S.C., Jamtveit, B. (2015). Continental crust beneath southeast Iceland. *PNAS*, 112(15), E1818–E1827.

Van Kreveld, S., Sarthein, M., Erlenkeuser, H., Grootes, P., Jung, S., Nadeau, M.J., Pflaumann, U., Voelker, A. (2000). Potential links between surging ice sheets, circulation changes, and the Dansgaard-Oeschger cycles in the Irminger Sea, 60-18 ka. *Paleoceanography*, 15, 425–442.

Van Vliet-Lanoë, B. and Guðmundsson, Á. (2020). Permafrost and climate change in Iceland. *Environ. Périglac.*, 2015–2018(22–23), 31–38.

Van Vliet-Lanoë, B., Bourgeois, O., Dauteuil, O., Embry, J.C., Guillou, H., Schneider, J.L. (2005). Deglaciation and volcano-seismic activity in Northern Iceland: Holocene and Early Eemian (The Syðra Formation). *Geodin. Acta*, 18, 81–100.

Van Vliet-Lanoë, B., Schneider, J.L., Guðmundsson, Á., Guillou, H., Nomade, S., Chazot, G., Liorziou, C., Guégan, S. (2018). Eemian estuarine record forced by glacio-isostasy (S Iceland): Link with Greenland and deep sea records. *Can. J. Earth Sci.*, 55(2), 154–171.

Van Vliet-Lanoë, B., Bergerat, F., Allemand, P., Innocent, C., Guillou, H., Cavailhes, T., Liorzou, C., Grandjean, P., Passot, S. (2020a). Tectonism and volcanism enhanced by deglaciation events in southern Iceland [Online]. *Quat. Res.*, 94, 94–120. Available at: http://doi.org/10.1017/QUA.2019.68.

Van Vliet-Lanoë, B., Knudsen, O., Guðmundsson, Á., Guillou, H., Chazot, G., Langlade, J., Liorzou, C., Nonnotte, N. (2020b). Volcanoes and climate: The triggering of Preboreal jökulhlaups in Iceland. *Int. J. Earth Sci.*, 109, 847–876 [Online]. Available at: http://doi.org/10.1007/s00531-020-01833-9.

Villemin, T. and Bergerat, F. (2013). From surface fault traces to a fault growth model: The Vogar fissure swarm of the Reykjanes Peninsula, Southwest Iceland. *J. Struct. Geol.*, 51, 38–51.

Voight, B., Clifton, A., Hjartarson, A., Steingrímsson, B., Brandsdóttir, B., Rodríguez, C., McGarvie, D., Sigmundsson, F., Ívarsson, G., Friðleifsson, G.O., Larsen, G., Jónsdóttir, G.S., Noll, H., McDougall, I., Kaldal, I., Friðleifsson, I.B., Aronson, J.L., Karson, J.A., Grönvold, K., Young, K.D., Kristjánsson, L., Sigurgeirsson, M.Á., Guðmundsson, M.T., Jancin, M., Flóvenz, Ó.G., Einarsson, P., Williams Jr., R.S., Buck, R., Pálmadóttir, S., Friedrich, W. (2020). A half-century of geologic and geothermic investigations in Iceland: The legacy of Kristján Sæmundsson. *J. Volcan. Geoth. Res.*, 391, 106434 [Online]. Available at: http://doi.org/10.1016/j.jvolgeores.2018.08.012.

Werner, R. and Schminke, H.-U. (1999). Englacial vs lacustrine origin of volcanic table mountains: Evidence from Iceland. *Bull. Volcan.*, 60, 335–354.

Wolfe., C.J., Bjarnason, I.T., Van Decar., J.C., Solomon, S.C. (1997). Seismic structure of the Iceland mantle plume. *Nature*, 245–247.

Zachos, J., Pagani, M., Sloan, L., Thomas, E., Billups, E. (2001). Trends, rhythms, and aberrations in global climate 65 Ma to present. *Science*, 292, 686–693.

Ziegler, P.A. (1992). North Sea rift system. *Tectonophysics*, 208, 55–75 [Online]. Available at: http://doi.org/10.1016/0040-1951(92)90336-5.

Websites

Geological maps, geothermal investigations (*Iceland Geosurvey, Ísor*): https://en.isor.is.

Glaciers: http://www.swisseduc.ch/glaciers.

Hydrography (Haffránsoknastoffnun, Marine and Freshwater Research Institute): https://sjora.hafro.is/.

Maps and aerial photographs of Iceland (*National Land Survey of Iceland*; *Langmælingar Íslands*): https://www.lmi.is/en/.

Maps of Iceland: http://map.is.

Meteorology, hydrology, seismology, volcanic activity (Icelandic Meteorological Office; Veðurstofa Íslands): https://en.vedur.is/.

Vatnajökull National Park: https://www.vatnajokulsthjodgardur.is/en.

List of Authors

Françoise BERGERAT
ISTeP
CNRS-Sorbonne Université
Paris
France

Laurent GEOFFROY
Géosciences Océan
IUEM
Université de Bretagne Occidentale
Brest
France

Hervé GUILLOU
Laboratoire des Sciences du Climat et de l'Environnement
LSCE/IPSL, CEA-CNRS-UVSQ
Université Paris-Saclay
Gif-sur-Yvette
France

René MAURY
Géosciences Océan
IUEM
Université de Bretagne Occidentale
Brest
France

Brigitte VAN VLIET-LANOË
Géosciences Océan
IUEM
Université de Bretagne Occidentale
Brest
France

Index

A, B

accretion, 6, 22, 42, 47, 119, 125, 130, 142–145, 149, 160, 163, 165–167, 169
aquifer, 37, 86
archives, 25
Arctic Ocean, 143
Bartonian, 18, 125, 145
basalt, 36
bathymetry, 115
bookshelf faulting, 82, 83
buoyancy, 123
Burdigalian, 125, 146

C

caldera, 27, 48
cap, 5, 14, 39, 43, 100
  ice, 5, 14, 39, 43, 100
central volcano, 27
CGPS, 23, 40
components, 7, 180, 182
contamination, 142
continental breakup, 6, 117, 118, 120, 127, 130, 132, 133, 136, 140–143, 147–149, 152, 155, 165, 169
convection, 92, 156, 157, 166, 167, 172, 175, 177
cooling, 56, 57, 142
crater, 73, 79
craton, 142, 156
crust
  continental, 11, 119, 121, 125, 130, 140, 142, 167–169
  oceanic, 8, 11, 115, 119, 121, 125, 130, 143, 152, 157, 159, 166, 169
crystals, 96

D

dating, 14, 15, 42, 43, 125, 135
  $^{40}Ar/^{39}Ar$, 43
  radiometric, 14, 43

decompression, 14, 119, 122, 124, 153
deglaciation, 13, 37, 38
diachronous, 149
diagram, 53, 80, 121, 135
discharge, 14, 38,
ductile, 142, 169
  flow, 121
dyke, 13, 15, 27, 38, 42, 48, 53, 54, 56, 63–66, 69, 90, 92, 103, 104, 119, 121, 127, 136, 140, 156, 175

E

earthquake, 24, 28, 30–32, 34, 36, 38, 72, 77, 80, 81, 96
Eemian, 62
en échelon, 35, 46, 48, 56, 61, 73, 78, 79
end, 75
englaciation, 14
Eocene, 16, 22, 125, 140, 142, 143, 152
esker, 38
eustatism, 14
EVZ, 13, 23, 27, 39, 40, 43–47, 51, 69, 70
exhumation, 119, 128, 168, 169
  passive, 169
extensive, 127

F

fault
  normal, 55, 56, 58
  seismic, 28, 30, 37, 75, 77
  strike-slip, 80
  zone, 7, 62
fissure, 48, 55, 97, 101, 103, 105, 107, 112
  swarm, 29, 48, 50, 52, 53, 62
flow, 93, 94, 151, 156
focal mechanism, 30, 31
fracture(s)
  open, 6, 9, 35, 49, 54, 55, 58, 63, 94, 96, 97, 99, 111, 131
  schistosity, 60
fracturing, 36, 50, 51, 55, 90, 92
fragile, 119, 160

G

gas, 14, 94, 96, 97
geoid, 115, 117
geothermal, 87, 88, 93
GIA, 39
GIFR, 1, 11, 14, 117, 148, 152, 157, 159, 160, 165–167, 169
GIR, 9, 16, 151, 154
glacial advance, 37
glaciers, 1, 5, 11, 13–15, 17, 24, 25, 39, 151
GPS, 24, 27, 32, 35, 38–40, 49, 98, 102, 103, 106–108, 111
graben, 50, 52, 73, 105
gravitational, 123, 132, 157
growth, 47, 54, 55, 57, 58

H

heat flow, 86
Holocene, 48, 69
hot spot, 6–11, 13, 14, 21–23, 45, 86, 123, 124, 151, 152, 155, 157, 158, 165, 166
hydraulic breccias, 60
hydrostatic pressure, 94

I

ice sheet, 136
IFR, 9, 151, 154

incompatible elements, 122, 155, 167
INSAR, 24, 32, 35
insularization, 1, 14
intrusion, 25, 108, 121, 140, 141, 160, 170
isochrones, 162

K, L

kaolinite, 94
lava
aa, 48, 77, 78
corrugated (*pahoehoe*), 77
flows, 14, 41, 43, 48, 54, 56, 77, 121, 134, 139, 140, 159
lake, 13
LBA, 2
linkage, 58, 148
LIP, 8, 118, 121, 122, 124
litosphere, 6, 8, 115, 117–119, 122–125, 130–132, 142, 149, 151, 155–157, 165, 167, 171

M

mafic, 8, 119, 121, 122, 125, 127, 134, 162
magma (*see also* reservoir), 8, 11, 38, 54, 103, 105, 108, 110–112, 119, 121, 123, 127, 136, 140–142, 152, 156, 157, 162, 165, 167, 169
magmatic province, 136, 139, 152, 155–157
magnetochrone, 125
magnitude, 13, 24, 27–32, 38, 67, 80, 81, 99
mantellic, 8, 21, 118, 119, 121, 123, 136, 140, 151, 152, 155, 157, 166, 167
mantle plume, 8–10, 21, 41, 123, 124, 151, 152, 155, 157
mantles (*see also* melting), 6, 8, 10, 14, 17, 115, 119, 122, 123, 128, 132, 151–157, 159, 165, 166, 169, 171
MAR, 6, 7, 12, 13, 47
maturity, 59, 84, 85
melting
adiabatic, 8, 10, 14, 38, 115, 118, 121, 123, 124, 136, 140, 152, 155–157, 159, 165, 167
mantle, 8, 10, 14, 38, 115, 118, 121, 123, 124, 136, 140, 152, 155–157, 159, 165, 167
partial, 8, 10, 14, 38, 115, 118, 121, 123, 124, 136, 140, 152, 155–157, 159, 165, 167
Messinian, 125
microseismicity, 13, 24, 27, 29, 30, 69
Miocene, 53
MORB, 122, 153

N, O

NADW, 15
Neogene, 1
North-Atlantic Current, 1
NVZ, 12, 23, 26, 27, 40, 42–46, 51, 59–61, 63, 68, 85, 96, 97, 164
obliquity, 147
oceanic propagator, 157
oceanization, 127, 130, 140, 142, 147, 155
Oligocene, 1, 17, 22, 125, 130, 143, 145, 149
outgassing, 140

P

pahoehoe, 48, 77
Paleocene, 125, 130, 133, 136, 140, 143, 151, 152, 155
paleo-rift, 22, 41–43, 45, 57, 148
passive margin, 121, 127, 129, 133, 141, 147, 149, 168, 169
peat bog, 14
picrite, 140, 154, 174
plate divergence, 42, 83
platform, 15–18
Pleistocene, 61, 62, 69, 72
Pliocene, 53
polarity, 125
propagation, 13, 23, 26–28, 47, 54, 56–58, 68, 108, 147, 157
pull-apart, 62, 64
pulse, 41
pumice, 73
push-up, 35, 75, 77–81, 97

Q, R

Quaternary, 13, 17, 36, 38
radiocarbon, 14
regression, 80, 81
reservoir, 94, 155
  magma, 94
rheology, 119
ridge, 6, 7, 12, 21, 22, 28, 30, 41, 46, 66, 85, 86, 117, 124, 127, 144, 146–149, 151, 152, 157, 166, 167
  hyaloclastite, 1, 6, 7, 9, 11–13, 16, 21–23, 44, 59, 69, 143, 147, 148, 150–154, 157, 159, 160, 162, 165, 166, 168, 169
  Mid-Atlantic, 1, 6, 7, 9, 11–13, 16, 21–23, 44, 59, 69, 143, 147, 148, 150–154, 157, 159, 160, 162, 165, 166, 168, 169
Riedel, 27, 82–84, 108, 109
rift
  active, 7, 10, 12–14, 20–24, 36–43, 45–48, 51, 53–55, 57, 58, 62, 69, 72, 86, 97, 99, 101–105, 108, 110, 111, 129, 131–133, 140, 147, 151, 162, 163, 165, 166, 169, 174, 177, 178
  jump, 7, 10, 12–14, 20–24, 36–43, 45–48, 51, 53–55, 57, 58, 62, 69, 72, 86, 97, 99, 101–105, 108, 110, 111, 129, 131–133, 140, 147, 151, 162, 163, 165, 166, 169, 174, 177, 178
rifting, 30, 48, 99, 100, 108, 109, 111, 112, 149, 170, 174, 176, 177
roll-over, 121

S

SDRs, 119, 121, 123, 125, 136, 140, 142, 159, 161–163, 165, 167–169
sea level, 16
sediment, 14, 61, 115, 119, 120, 138
segmentation, 75, 99
seismic
  activity, 10, 12–14, 22, 24, 25, 28, 30, 32, 35, 37, 38, 44, 60, 61, 69, 70, 74, 76, 81, 98, 99, 119, 121–123, 129, 140, 151, 153, 159, 160, 162, 169

crisis, 10, 12–14, 22, 24, 25, 28, 30, 32, 35, 37, 38, 44, 60, 61, 69, 70, 74, 76, 81, 98, 99, 119, 121–123, 129, 140, 151, 153, 159, 160, 162, 169
fault, 10, 12–14, 22, 24, 25, 28, 30, 32, 35, 37, 38, 44, 60, 61, 69, 70, 74, 76, 81, 98, 99, 119, 121–123, 129, 140, 151, 153, 159, 160, 162, 169
reflection, 14, 60, 61, 119, 122, 123, 159, 169
tomography, 123, 151
seismogenic, 160
Serravallian, 125
shear, 36, 38, 52, 53, 55, 56, 66, 82–85, 119, 156, 167
sill, 13, 121, 140
SISZ, 12, 13, 22, 24, 31, 32, 38, 44, 69, 70, 73–78, 81–85
solfatare, 97
stage, 58, 59, 80, 84, 140, 167, 168
strandflat, 5, 15, 17
stress, 55, 62, 65, 67, 68, 82, 84–86, 127
subduction, 8, 157, 176
subglacial, 2, 38, 45
subsidence, 18, 27, 105, 115, 117, 128
supercritical (water), 94
suture, 1, 10, 131
swarm, 45
swelling, 27
syn-rift, 119

T

tectonic plate, 7, 130
tectonics, 1, 8, 13, 14, 18, 27, 36, 39, 40, 44, 48, 55, 65, 115, 119, 121, 122, 132, 141, 143, 156, 159, 162, 166
TFZ, 12, 13, 22, 24, 27, 28, 44, 59–61, 63, 67, 68, 84
thermal
contraction, 115
relaxation, 132, 142
thermohaline circulation (THC), 1, 11, 16, 18, 125
thinning, 39, 118, 119, 123, 128, 132, 133, 140, 169
crustal, 133
Thulean magmatism, 133, 135, 151
Tortonian, 18
transtension, 62, 130, 132, 140
trap, 8, 121, 123, 124, 133, 134, 136, 140, 147, 155

U, V, W

underplating, 152
upwelling, 117
vapor, 88, 94, 96, 97
volcanic cone, 72
WVZ, 12, 13, 23, 39, 40, 43–48, 53, 69, 70, 72, 88, 94

Summary of Volume 2

List of Abbreviations

Preface
Brigitte VAN VLIET-LANOË and Françoise BERGERAT

Introduction
Brigitte VAN VLIET-LANOË and René MAURY

Chapter 1. Young Icelandic Volcanism and its Implications
René MAURY and Brigitte VAN VLIET-LANOË

1.1. Introduction
1.2. Icelandic magma series
1.2.1. Lava types
1.2.2. Geochemical diversity of young Icelandic basalts and their sources
1.2.3. Some geochemical constraints concerning the origin and geodynamic evolution of Iceland
1.3. Central volcanoes and active fissural systems
1.3.1. Central volcanoes
1.3.2. Fissural volcanism and subaerial lava flows
1.3.3. Hydromagmatism
1.4. Volcanic hazards in Iceland
1.4.1. Hazards related to lava flows
1.4.2. Hazards related to explosions and gas emissions

1.4.3. Jökulhlaups and associated hazards
1.4.4. Icelandic dust: a consequence of volcanism
1.5. References

Chapitre 2. Volcanism and Glaciations: Forcings and Chronometers
Hervé GUILLOU, René MAURY and Brigitte VAN VLIET-LANOË

2.1. Subglacial volcanic landforms
2.1.1. Subglacial isolated volcanoes or tuyas
2.1.2. Hyaloclastite ridges or *tindar*
2.2. Volcanism, deglaciation and climate
2.2.1. General features: deglaciation, discharge and partial melting
2.2.2. Deglaciation and climate feedback
2.3. The hypothesis of a link between volcanism and climate and its test by dating
2.3.1. The K-Ar chronometer
2.3.2. The combination of K-Ar and $^{40}Ar/^{39}Ar$ methods for dating Icelandic volcanism
2.3.3. A link between volcanism and climate according to K-Ar ages?
2.3.4. A rhyolitic volcanism synchronous with deglaciations?
2.4. References

Chapitre 3. Cenozoic Evolution of Iceland and the Cryosphere
Brigitte VAN VLIET-LANOË and Hervé GUILLOU

3.1. Ice ages and the opening of the Atlantic
3.1.1. The Middle and Final Miocene cooling
3.1.2. The acceleration of the Middle Pliocene
3.1.3. The Middle Pleistocene transition
3.1.4. The initiation of thermohaline circulation
3.2. Iceland's Quaternary glaciations
3.2.1. Conditions for the development and functioning of ice caps
3.2.2. Glacio-isostasy
3.2.3. Icelandic data
3.2.4. The Icelandic record
3.3. The last glacial episode and its deglaciation
3.3.1. The Weichselian
3.3.2. The Last Glacial Maximum
3.3.3. Deglaciation and the Holocene

3.4. Iceland today, its climate and vegetation
3.4.1. The climate
3.4.2. Ocean circulation and climate
3.4.3. Soil, people and climate
3.4.4. Soils and erosion
3.5. References

Conclusion
Brigitte VAN VLIET-LANOË, René MAURY and Hervé GUILLOU

References